T0138276

The Halle Orphanage as Scientific Community

The Halle Orphanage as Scientific Community

Observation, Eclecticism, and Pietism in the Early Enlightenment

KELLY JOAN WHITMER

The University of Chicago Press   Chicago and London

KELLY JOAN WHITMER is assistant professor of history at
Sewanee: The University of the South.

The University of Chicago Press, Chicago 60637
The University of Chicago Press, Ltd., London
© 2015 by The University of Chicago
All rights reserved. Published 2015.
Printed in the United States of America

24 23 22 21 20 19 18 17 16 15   1 2 3 4 5

ISBN-13: 978-0-226-24377-1 (cloth)
ISBN-13: 978-0-226-24380-1 (e-book)
DOI: 10.7208/chicago/9780226243801.001.0001

Library of Congress Cataloging-in-Publication Data

Whitmer, Kelly Joan, author.
  The Halle Orphanage as scientific community: observation,
eclecticism, and pietism in the early Enlightenment / Kelly Joan
Whitmer.
     pages; cm
  Includes bibliographical references and index.
  ISBN 978-0-226-24377-1 (cloth: alk. paper) — ISBN 978-0-226-24380-1
(e-book) 1. Waisenhaus (Halle an der Saale, Germany)—History—
18th century. 2. Science—Study and teaching—Germany—Halle an
der Saale—History—18th century. 3. Science—Germany—Halle an
der Saale—History—18th century. 4. Education—Germany—Halle
an der Saale—History—18th century. 5. Observation (Scientific
method)—Germany—Halle an der Saale—History—18th century.
6. Eclecticism—Germany—Halle an der Saale—History—18th
century. 7. Pietism—Germany—Halle an der Saale—History—18th
century. I. Title.
  Q183.4.G32H35 2015
  506.043'1848—dc23
                          2014029026

To Charles and all of my parents

Unity of faith must flow out of wisdom,

And *wisdom* is building a house of *unity*, listen;

Those who would like to know something about the large structure:

The wise and wonderful God, who alone commands praise,

Directs me constantly toward true wisdom,

And is building, as has long been planned,

A huge house of virtue and schools of true wisdom:

In this same work, many boys allow themselves to be taught wisdom

Through the arts and sciences,

To God's praise and honor and their parents' reputation and happiness,

Although not every son, without difference, can study there.

Only the best ones are sent,

So that they will immediately begin to blossom and begin to love,

Until the spirit of wisdom begins to peek out of his soul,

And raw utility becomes corporally actualized.

In this school of wisdom one only takes in boys,

Who, by nature are oriented toward knowledge,

The parents can even be Jews and Turks,

It is true, Protestants attend there alongside Papists,

The children do not hear about even the slightest hint of confessional strife,

But instead learn virtues and the arts from all sides,

So that they grow in years and in wisdom,

And this means they have to work, themselves, on hundreds of things.

The same kind of project is to be seen nowhere else,

And one will find much more there than what one can speak of here,

The Orphanage in Halle is truly good and fine,

There is really nothing that can be counted against it.

JOHANN ERNST BESSLER, *DER RECHTGLAUBIGE ORFFYREER*

Most people would agree that learning mathematics makes other studies easier and go more quickly than other disciplines that do not deal with logic, the art of meditation and demonstrations. . . . After mathematics, in schools it is also important to teach physics: a science of nature, including the essences, qualities, powers and effects of material things in the world, best apprehended through experience, also observations and experiments.

. . . Experience proves that the Latin young people learn in schools rarely helps them once they leave. . . . Better would be to focus on learning more useful languages, like German, which is often neglected, French and Italian, or English. Geography and history are also very useful subjects to teach in schools. . . .

I like the *Pädagogium* and the Orphanage better than many other schools because there the youth are directed more toward real things and forms of knowledge that exercise their understanding than things that only confuse the memory. Not to mention that they are far ahead of other schools in the way that they provide instruction in Latin, German and the gallant languages and arts: by setting a good example and offering basic directions, they give young people practical opportunities to practice their morality and to grasp the mathematical and physical sciences through many experiments. FRIEDRICH HOFFMANN, "VORREDE"

It would be good if the observer [der Observator] were not only a mathematician, but also a student of nature, who observed plants, animals, minerals and other naturalia and artificialia in various locations, because this all goes together and can be done at the same cost. The minerals will be useful for paving the way for new mines, the plants and animals for commerce and manufacturing, everything will contribute to the further development of physics. . . .

He must travel with special instruments. . . . What is important is a good odometer, compasses, quadrants and similarly made instruments, large pendulum clocks, levels, optical instruments, microscopes, micrometers, barometers, good magnets, magnetic globes. Especially useful would be an Instrumentum inclinatorium, so that magnetic inclination, which is different from declination, could be observed. GOTTFRIED WILHELM LEIBNIZ, "CONCEPT EINER DENKSCHRIFT"

Contents

ONE

Introducing the Orphanage

In the middle of the 1690s, a group of German Lutherans known as Pietists founded an orphanage in the Prussian city of Halle (an der Saale). The building (fig. 1) in which it was initially housed, known simply as "the Halle Orphanage" (*das hallesche Waisenhaus*), was actually the first in a series of buildings that became a much larger suburban institutional facility. Over time Halle's Orphanage came to function as a "showplace" (*Schauplatz*), a place worth contemplating and observing. It was the headquarters of an educational, charitable, and scientific community comprising an elite school, or *Pädagogium*, for the sons of noblemen; schools for the sons of artisans, soldiers, and clergy; a hospital; an apothecary; a bookshop; a botanical garden; and a gallery of architectural models, *naturalia*, and scientific instruments. The entire facility, including the curious objects it housed, attracted thousands of visitors such as those shown in figure 2 in front of the main building. Standing mostly with their backs to us so that we cannot see the expressions on their faces, their postures suggest that they are captivated by what they are seeing. Two of the figures stand with their hands on their hips, lingering for a while to take everything in. Another figure, clearly engaged in conversation, gestures at the building with his cane.

Many reported being profoundly inspired upon seeing the Halle Orphanage (hereafter referred to as the Orphanage) for the first time. When he first observed it upon assuming the throne of Brandenburg-Prussia in 1713, Friedrich Wilhelm I, reported he had marveled at its enormous size.[1] Others indicated that the effect of observing the Orphanage

1 David Ulrich Boecklin, *The Halle Orphanage* [1730], engraving. Credit: Halle (Saale) Stadtarchiv.

had been so intense that it inspired them to found similar institutions in their own hometowns. And indeed, many did build replica orphanages in east Brandenburg (Neumark) and in Saxony, Bavaria, Denmark, Siberia, southern India, and even North America.

Because of its size and scale, the Orphanage has long held a prominent place in histories of Pietism, Brandenburg-Prussia, and early modern central Europe. It was a site of innovation in medicine, including the preparation of pharmaceuticals, and served as the headquarters of the world's first Protestant mission to India.[2] Scholars have long recognized that there was a great deal going on inside this space, as well as in the city of Halle and its university by 1700. However, because of its reputation as a Pietist enclave

2   Detail from the Boecklin engraving (fig. 1).

inhabited largely by young people, hardly the gentleman savants associated with the period's scientific academies and societies, the Orphanage has not been taken seriously as a scientific community. Yet this is precisely what it was. Inside this institutional ensemble and spaces modeled after it, student teachers, their pupils, pastors, and professors worked together to interrogate natural processes through tests and by refining a range of observational procedures.

Inside the Orphanage

The first director of the Halle Orphanage was a controversial professor of Greek, oriental languages, and, later, theology at the University of Halle named August Hermann Francke. In 1701, he published a report detailing his plans titled "Project for a universal seminar or the creation of a garden, in which one anticipates the real improvement of all social orders both in and outside of Germany, indeed in Europe and in all other parts of the world."[3] In another description of the project from 1711, Francke explained that from the very beginning, the Orphanage was to be closely connected to the University of Halle (founded 1694). "The purpose," he wrote, "is to build a universal facility [*eine Universal-Einrichtung*] near the university for the use of Christendom and the entire world."[4]

The facility was to comprise nine institutions, "some of which have been realized," he continued, "some of which have not yet been founded."[5] These institutions included

1. The Orphanage [*das Waisenhaus*], inside of which approximately 100 young boys and girls can be found and which is made up of two apartments, a big one and a small one, and a special part of the house for girls younger than two years old.
2. The schools of the Orphanage, in which at last count there were 1,333 pupils (including 133 orphan children) and 80 teachers, who work daily with the youth.
3. An advanced boarding school [*Pädagogium*] for children of people who are well

off and of a certain status, now 70 scholars—most of whom are foreigners—and 12 teachers. . . .

4. A seminar for teachers [*Seminarium Praeceptorum*]
5. An institution for women: 1) for adult noble and bourgeois women who live in their own foundation partly from their own funds and partly from bequests; 2) for the young daughters of noble and bourgeois people, who are raised here and instructed in different kinds of feminine work . . . at their own cost; 3) for poor widows. This three-part institution exists and is divided into three apartments that are entirely separate from each other; and it is good that they remain so divided. . . .
6. An infirmary or institution for sick and weak people
7. A workhouse
8. A college of oriental studies [*Collegium orientale*]
9. A seminar of nations [*Seminarium Nationum*][6]

The purpose of the *Collegium orientale*, founded in 1702, was to offer instruction in the Chaldean, Syriac, Arabic, and Ethiopian languages; in Rabbinic or Talmudic writings; and in ancient Greek, Hebrew, Latin, and, when possible, Armenian, Persian, Chinese, Turkish, Greek, Polish, Slavonic, and Russian.[7]

The "seminar of nations" did not yet exist in 1711, but Francke noted he was looking for ways to inspire several "foreign nations" to send their children to Halle to be raised and educated. As early as 1701, he had taken in four boys from Silesia, one from Prague, and one from Holland; international contacts continued to send young people to be educated in Halle from as far away as Moscow, Sweden, and Denmark.[8] Four thirteen-year-old boys arrived from London on December 12, 1706.[9] They learned alongside young men who had traveled to the "universal facility" from virtually every city in the Holy Roman Empire: Leipzig, Dresden, Berlin, Magdeburg, Hamburg, Heidelberg, Frankfurt am Main, Gotha, Erfurt, Weimar, Zörbig, and more. When Francke died in 1727, there were 2,200 children attending school at the Orphanage; of these 100 were orphan boys and 34 were orphan girls, and 170 student teachers were living and working in the community.[10]

Now let us take a tour of the facilities alongside King Friedrich Wilhelm I, who arrived for a visit on October 4, 1720.[11] Francke met him in front of the main building (fig. 3), led him up the steps and inside, where he showed the king the apothecary, the bookstore, and sleeping quarters.[12] These were housed in the upstairs apartment, which later became the site of the Orphanage's museum. After climbing to the top floor and looking around, Francke and the king walked down the steps and exited through the back of the building. They turned right and walked across the covered part of the inner courtyard, where they entered an adjacent

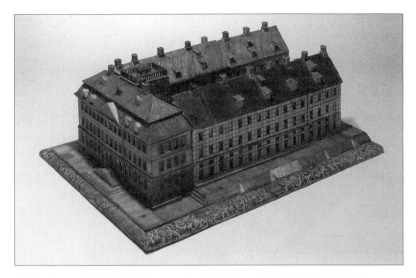

3   Wooden model of Halle's Orphanage. Credit: Archiv und Bibliothek der Franckesche Stiftungen [AFST/B A0776].

building (fig. 4). It contained an auditorium and a cafeteria; the tables are visible in the model.

After discussing the costs of feeding boarders and touring the kitchen, Francke and the king returned to the courtyard, where they stood in front of the row of buildings immediately across from them, parallel to the wing they had just visited. The king reportedly asked, "Who lives in here?" and Francke told him that these apartments housed the inspector of the *Pädagogium* and other schools in the complex. They also housed students from the University of Halle, many of whom taught in the schools in exchange for their rooms, and candidates for the Lutheran ministry. There were also several rooms where eighteen to twenty male pupils of the Orphanage's schools lived with a teacher as part of a small household. Those who could paid for their room and board; those without means were not expected to pay.[13]

The king then inquired about the community's schools and learned that there were 14 classes of pupils (621 in total) who attended the vernacular or "German school." The more talented pupils had been grouped into eight classes and were taught by thirty student teachers in the Latin school. Pupils in the Latin school were instructed in theology, Latin, Greek, Hebrew, history, geography, geometry, writing, and arithmetic. Those attending the German school "received free paper and books" and learned the catechism, reading, writing, arithmetic, geography, history, and some Latin.[14]

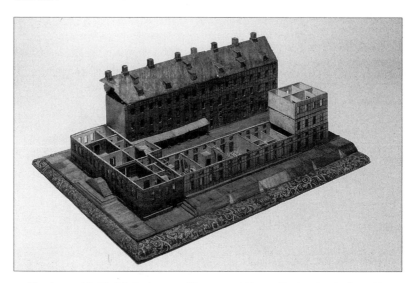

4   Wooden model of Halle's Orphanage with top removed, revealing interior. Credit: Archiv und Bibliothek der Franckesche Stiftungen [AFST/B A0909].

As they continued to talk about the schools of the Orphanage, the two men slowly walked up a small hill to the opposite side of the campus to the *Pädagogium Regium* (fig. 5). The king asked if and how the *Pädagogium* differed from the other facilities.[15] Francke explained that it had its own school and living quarters for young men from wealthy and titled families. They toured the building, including the mechanical room, where there were ten wood-turning stations (*Drechselbanke*) in which three pupils per station took turns showing the king their "exercises." According to a future *Pädagogium* director's description of these exercises, they involved "crafting wooden models" using a variety of woodworking instruments. "The purpose of these exercises," Hieronymous Freyer wrote, "was not to produce a variety of finished products but to generate motion and knowledge [*Wissenschaft*]."[16] After letting him observe the exercises, several teachers and pupils performed an experiment with the air pump, which the king "observed with great contentment."[17]

Next Francke took the king back into the inner courtyard between the main Orphanage building and the *Pädagogium*, where all the community's young people had gathered and were standing in rows. The king wandered among these children and talked "especially to those whose clothing indicated that they were the sons of soldiers."[18] These children had to step forward and tell the king the name of their father and their father's company. The king then asked what else there was to see. Francke replied

5 *Prospect des Innern Waysenhauses* [ca. 1750]: a view of the Orphanage ensemble's interior with *Pädagogium* in the distance, engraving. Credit: Archiv und Bibliothek der Franckesche Stiftungen [AFST/B Sa 0014].

that a mechanical room had recently been built inside the main building, and after his experience in the *Pädagogium*, the king insisted upon seeing it. On his way there, the king was taken to the Orphanage's library, which had not yet been moved to its own building. Next to the library, the king saw a wooden model of the Holy Land and asked how one knew that it had really looked that way in the past. Someone answered that the model "had been built using the best descriptions and maps available."[19] Among the other curiosities he saw once back inside was an instrument for measuring longitude, which he was told had been invented by a local preacher and mathematician named Christoph Semler, who had allowed it to be carried to England.[20] By this time it was getting dark; the king made his way back out the main entrance, thanked Francke, and took his leave.

Pietism, Philanthropy, Enlightenment

Francke was well known in his own day not only as the director of Halle's Orphanage but also as the leader of a spiritual renewal movement called Pietism. Scholars still debate whether to define Pietism as a response to a general crisis of piety affecting Protestants and Catholics alike in the seventeenth century or as mainly a response to Lutheran orthodoxy.[21] Exploring strategies for generating love and communal solidarity and for renewing and intensifying expressions of religious feeling became one of the most important activities of Pietist groups, including "Halle Pietists" and the communities of "radical Pietists" that fall outside the scope of this study.[22] So in many ways it makes sense to apply the term broadly and even loosely. At the same time, there were very specific circumstances surrounding the original use of the term that are particularly important to keep in mind, especially when focusing on the Halle Pietists.[23] When orthodox theologians first began using the word *Pietist* in Leipzig in the 1680s and 1690s, they did so in order to single out, even to insult, those who adopted an antagonistic position toward the orthodox Lutheran establishment.[24] Pietists were considered dangerous because they asked too many questions, disrespected authorities, and seemed to be heading in the direction of religious enthusiasm.

In his youth, Francke acquired a reputation as a pot-stirrer in Leipzig, so it is little wonder that his critics later associated him and other Pietists with enthusiasm. This is a term that could mean many things too: many of those labeled enthusiasts, or *Schwärmer*, claimed to receive divine messages they said encouraged them to challenge the existing sociopolitical order. Their critics were quick to label them as antisocial and irrational.[25] Francke did not directly challenge the social order, but he questioned the relevance of philosophy and logic for the study of theology.[26] Since he was working with theology students at the University of Leipzig, orthodox theologians there feared he was undermining their authority; they conducted an official investigation and prohibited him from lecturing. To make matters worse, when Francke left Leipzig and moved to Erfurt, he made it known that he had visited and approved of some local female prophets and their epiphanies; he was eventually deemed too radical by civic authorities and forced out.[27]

Once his mentor, Philipp Jakob Spener, moved to Berlin and got Francke involved in state-mandated efforts to start a new university in Halle, both men realized that Francke was going to have to be less outspoken. If the Orphanage was to be heavily subsidized by and associated with Prussian leaders, then it could not be a source of disorderly forms of behavior. Instead its founding marked an opportunity to study enthusiasm, especially the interplay among affect, intelligence, and moments of inspiration, which were much less disruptive (and easier to regulate) than spontaneous outpourings of emotion. Throughout his tenure as director, Francke preferred to avoid using the term *Pietist* altogether—it was too politically loaded. Those involved with the Orphanage usually referred to themselves as *Hallenser*.[28] I have generally followed their lead in this, preferring to discuss specific practices, objects, and contributions by individuals associated with the Orphanage rather than using the term *Pietist* when it does not appear in my sources.

Clearly, the Orphanage's affiliation with the Prussian state is key to understanding how it functioned as a community, and the work of Klaus Depperman and Carl Hinrichs made it clear long ago that this relationship was worth a closer look.[29] Since then, several historians have viewed the Halle Orphanage as a place to learn more about the modernizing strategies used by heads of territorial states, who were trying to consolidate political boundaries—and their own power—in this period.[30] My study of the Orphanage is indebted to these efforts, including the work of Anthony La Vopa, who has written about the professionalization of "poor students" during this period of state expansion and consolidation. Thanks largely to the support of patrons such as Francke who were helping territorial leaders construct a meritocracy, young men who might not have been able to attend university could do so in exchange for their service as teachers.[31] Their status as poor yet gifted students was important for cultivating the Orphanage's identity as a philanthropic organization. And these mostly nameless young men played crucial, yet hitherto unexamined, roles in ongoing efforts to create forms of scientific community in central Europe and beyond.

These efforts transpired alongside related efforts to articulate a new, hybrid model of charity, which helped lead to the bureaucratization of daily life and the creation of a "police state" in Prussia.[32] The organization's primary concern was *not* to provide relief to the local urban poor.[33] It was not a typical city orphanage intended to house legitimate children from local families.[34] When Friedrich Wilhelm I came to tour the Orphanage in October 1720, he was heeding the advice of individuals known as cameralists and "police scientists," who, as William Clark has explained, called for the state to provide "an insulated infrastructure of entrepreneurial activity"— whether that activity occurred in a university, a mine, or an orphanage.[35] The state, with the help of its ministers, oversaw the activities of entrepreneurs, who were given freedom of action and movement as long as what they were doing served the state in some way. These entrepreneurs were increasingly called upon to demonstrate to ministers of the state that they had the proper mixture of talent, training, and credentials to justify the kind of work they performed.[36] The "insulated infrastructure of entrepreneurial activity" increasingly provided by the Prussian state meant that the Orphanage was a "police institution" at the same time as it retained its entrepreneurial ethos and transregional agenda.[37] Like most things related to the Orphanage, its role was not black or white but often gray.

The blurry and even contradictory qualities of the Orphanage are crucial to understanding its identity as a "universal facility." In addition to combining the aspirations of a centralizing territorial state with those of individual interests, Francke and his colleagues became involved in a variety of conversations about reconciling confessional difference. Francke also chose to accept the advice of a variety of individuals, including Gottfried Wilhelm Leibniz. This does not mean he always agreed with these individuals on every point; however, the fact that he was having these conversations at all is noteworthy.

In figure 6, Christian Thomasius, who was a professor of law in Halle, is shown standing in front of the main university building talking with Francke, who appears in front of the Orphanage (on the left). The hands of both men are stretched toward one another; they appear to be deep in conversation. This engraving was produced after the deaths of both these figures and is included in a work presenting an imaginary conversation between them. Yet the topic of their discussion was actually less

6   Frontispiece showing August Hermann Francke (left) and Christian Thomasius (right) in conversation, from *Besonders curieuses Gespräch am Reich der Todten* [1729], engraving. Credit: Staatsbibliothek Berlin [39 in:@Schoepp. 804], courtesy of Art Resource, NY

important in this instance than *how* they were discussing it—in an atmosphere of mutual respect and engagement.

Like the Orphanage, the University of Halle was also a model community, one that promoted polite conduct, good manners, and fruitful conversations even (or rather, especially) among people who disagree. It represented an alternative way of thinking about the possibilities of academic life as a space in which to promote interconfessional dialogue and reconciliation. The first statutes explicitly stated that the university's main goals were to put an end to confessional polemics, to generate tolerance, and to promote harmony among the four faculties: theology, law, medicine, and philosophy.[38] In addition to these statutes, explains John Holloran in his foundational study, Halle's professors actively strived to cultivate "a new attitude toward academic controversy."[39] They attempted to undermine, or at the very least challenge, the lingering culture of disputation and constant infighting among colleagues from different faculties that had become a normal part of university culture.[40] By doing this, Halle's professors were trying to transform the university for the better by replacing an outmoded scholastic curriculum centered on intellectual jousting and personal grudges.

Each professor articulated competing visions of reform, though, and wrestled with how to understand the underlying causes of interpersonal conflicts. This led to intensified reflection on the role of affect, passions, and desires in decision making.[41] Those involved in these ongoing discussions generally understood love as a force that interacted with cognition and perception in powerful ways. Francke engaged in these discussions by writing and preaching about the epistemological dimensions of philanthropy, or the relationship between perception, cognition, and affection. He was in a unique position to help others understand the original meaning of the Greek word *philanthrōpos* and the Latin word *philanthropîa*. The root word *phileô* signifies "affectionate regard" or friendship and was one of four words used by the ancient Greeks to describe different forms of love (along with *eros*, *stergô*, and *agape*).[42] In 1705 Francke gave two sermons entitled *Philanthropia Dei*, later published by the Orphanage; the following year he wrote a short treatise on the meaning of the term *philotheïa* that explored the nature, causes, and effects of different kinds of love.[43]

Francke's colleague Thomasius worried that the growing acceptance of a mechanistic worldview minimized attention to the effects of affective forces, tendencies, or passions long believed to be located in the body—now rendered inert and lifeless by mechanical philosophers.[44] He believed affect actually had a determining effect on cognition. Leibniz disagreed with Thomasius and argued instead that "it is cognition or perception that determines the basic quality of our affects."[45] Generating love or philanthropy did not happen suddenly (through revelation or a mystical conversion experience, for example), he insisted; this took practice and a great deal of work.[46]

Halle's institutional culture was defined by a profound commitment to exploring the implications of these positions, in the interest of finding a middle ground. True, Leibniz was not a professor at the university, but there were others there—such as professor of medicine Friedrich Hoffmann—who were avowed proponents of mechanism and admirers of Leibniz.[47] Hoffmann's colleague in the faculty of medicine Georg Ernst Stahl developed his own body of medical theories intended to counteract mechanist accounts of human nature and behavior.[48] Francke engaged Stahl, Hoffmann, and Thomasius in conversation, and he reached out to Leibniz and his friend the mathematician Ehrenfried Walther von Tschirnhaus, both of whom visited the Orphanage in person.[49]

I have chosen to delve into the story of Leibniz's and Tschirnhaus's interests in the Orphanage to illustrate just how seriously these scholars took its potential to serve as a scientific community. This is a particularly important intervention to make because of a lingering tendency to treat the space as an exclusively Pietist enclave and to see Pietists as "enthusiastically" opposed to rational approaches to knowing the natural world—and to science and the Enlightenment more generally.[50] Such a tendency is linked to an older generation of Enlightenment scholarship—generally associated with the work of Peter Gay—that treated the Enlightenment as a secular movement opposed to all forms of religion.[51] Related to this were assertions that "science" and "religion" were oppositional categories—science *versus* religion—even though "science" did not become a unitary category in Europe until well after the moment under consideration here. The present work is indebted to scholarship that has clarified just how deeply intertwined natural philosophical and theological concerns were in early modern Europe, even in the eighteenth century.[52]

Scholars now more frequently define the Enlightenment as "a series of interlocking, and sometimes warring problems and debates" about the origins and technologies of power, rationality, universal truths, and tolerance that unfolded throughout the eighteenth century in a variety of contexts.[53] It no longer makes sense to argue Pietists adopted antagonistic stances toward the Enlightenment, because it was not a stable or singular phenomenon—neither was Pietism. Yet at the same time, it is hard to ignore the fact that a group of theologians in Halle banded together to have the philosopher Christian Wolff forcibly removed from the city in 1723. Despite the ways in which the Enlightenment has been redefined, in other words, a binary Pietism-versus-Enlightenment framework lingers thanks in large part to Wolff's banishment.[54]

Traditional accounts of the Wolff expulsion usually point to a speech he gave titled "On the Practical Philosophy of the Chinese" during a ceremony marking the end of his term as university rector in 1721.[55] There was nothing in the speech that Wolff had not said before in his writings or his lectures, and his ideas were consistent with Leibniz's comments about the power of reason to determine the quality of our affections and philanthropy. As I will show in this book, these were ideas that Francke and others remained open to in the years leading up to Wolff's expulsion. Yet the speech itself became the turning point in a notorious controversy involving several members of Halle's theology faculty, who used their connections in the court of Friedrich Wilhelm I to have Wolff banned from the city.

Public controversies have long been important subjects of study because of the special insight they often provide into key issues and politics animating communities.[56] Making sense of the Wolff controversy certainly requires some focus on hot topics that were making waves in the city, but it also requires some consideration of the social reality it was situated within.[57] My goal is not to offer a new explanation for why Wolff was kicked out of Halle but to use my study of the Orphanage and its schools to draw attention to some of the issues at stake. One key issue involved how to understand the relationship between the faculties of theology and philosophy at German universities—a relationship that had grown exceedingly tenuous by the end of the eighteenth century.[58] A related issue concerned Halle-based theologians' close ties to courtly circles in Berlin and the power they had to regulate access to scientific methods and technologies.

A key Orphanage administrator named Johann Daniel Herrnschmidt, who was second in command after Francke by 1716, acknowledged that despite his conviction that everyone *ought* to observe, to study, and to reflect upon the natural world, not everyone has the time, opportunities, occupation, or inclination to investigate "physical things" in their "philosophical context" (*Philosophischen Zusammenhang*)—or to construct "rational arguments" about these things.[59] It required the right constellation of circumstances and a great deal of work to be in a position where one might begin to develop explanatory frameworks or theories. This is where the teacher came in, carefully selecting gifted pupils who had the potential to gather up observations and perform experiments. Acting as a sage or judge, a teacher assessed pupils abilities to pursue their own research by evaluating their behavior and performance. Orphanage student teachers had frequent conferences in which they discussed the progress of their pupils and made collectively reinforced statements about character, talent, and potential.[60] These collective assessments provided a means of controlling who did and did not get to do the theorizing, synthesizing, and explaining.

Were these theologians well suited to directing ongoing efforts to standardize training in experimental and observational methodologies, including deciding who had the right to develop frameworks and theories? Many believed the answer was yes.[61] After all, theology was still regarded as the highest of all the faculties. Yet at other German universities, at the University of Helmstedt, for example, it was not the theologians but the members of the philosophy faculty who took up the task of introducing experimental and observational procedures to the study of physics in the interest of devising new theories or explanations. These professors drew attention to the need to conduct research with the microscope and other instruments to improve the study of anatomy and natural history, for example.[62] This was also the case at the University of Altdorf, where Johann Christoph Sturm—a professor of mathematics and physics, not theology—started a collegium devoted to performing experiments in 1679.

As a mathematician affiliated with the university's philosophy faculty, Wolff possessed a special kind of expertise in mathematics and experimental physics that Halle's theologians did not have. Following the lead of his colleagues at other institutions, he offered a *Collegium physicum experimentale* at the university as early as 1707.[63] He even wrote a *Beginner's Guide to the Mathematical Sciences* that student teachers used in the Orphanage.[64] Wolff also believed he had a right—even an obligation—to theorize, "to philosophize," and to create universal frameworks.[65] His efforts to lay a foundation for his own system of philosophy in his *Rational Thoughts about God, the World, the Human Soul and Everything Else* (1720) by showcasing the power of mathematics wedded to philosophy led him to venture too far into terrain Halle's theologians had claimed for themselves.[66] They may have feared that he aspired to take over the Orphanage as well.

Eclectic Observers

At a pivotal moment in the emergence of a public sphere in central Europe, I contend that the Halle Orphanage emerged as a key "scientific community."[67] But what does this actually mean? Historians of science have long pointed to the emergence of pan-European scientific societies, academies, and associations in the seventeenth and eighteenth centuries whose members worked together to standardize a variety of empirical methods and technologies for assimilating information and generating a more material, or hands-on, knowledge of nature.[68] Sponsored by territorial leaders, these new associations were generally made up of noblemen-scholars and upwardly mobile members of Europe's professional classes. In recent decades, social and cultural historians of science have examined the forms of scientific sociability, or civility, pursued in these communities.[69] They have also explored other places and opportunities for individuals to manipulate materials and cultivate forms of expertise.[70] Historians of the Holy Roman Empire have paid special attention to the importance of alchemical research in this regard, including efforts to bring artisanal skills and knowledge of the material world under the control of territorial sovereigns.[71] These efforts have contributed to a recalibration of conventional accounts of the origins of modern science, which involved an astounding array of efforts to reconfigure relations between art and nature and collaborations between craftsmen, scholars, and the state. Making a case for the Halle Orphanage as a "scientific community" fits entirely within a much broader and sustained scholarly effort to draw attention to the diversity of venues and practices that might qualify as "scientific" in the early modern period. Given the evidence of all this diversity, then, was the Orphanage really so unique?

To be sure, the founders of Halle's Orphanage were engaged in a transnational and transgenerational discourse about the reform of the sciences and how to teach these sciences in schools. Before its founding there had been no shortage of utopian, reform-minded plans for improvement in virtually every corner of Europe and among Protestants and Catholics alike.[72] The Czech educator and natural philosopher Johann Amos Comenius was a major source of ideas.[73] During and after the Thirty Years' War (1618–48) he had advocated standardizing empirical methods by introducing object-based pedagogies in schools. One of Comenius's teachers, Johann Heinrich Alsted, had advocated a similar set of educational reform efforts focused on using objects to help improve the memories of young people so they would be able to see the interconnectedness of all things.[74] Reinvigorated schools would become repositories—ideally with museums that would contain the world in a single room—and research facilities. These efforts did lead to some early educational reforms, to the prolific expansion of schools run by Jesuits, and to the creation of new academies and societies.[75]

But in central European cities, scientific societies were few and far between; a scholastic-inspired curriculum still dominated most schools.[76] By 1700, central European universities were only just beginning to be imagined as spaces for scientific research, and there were still very few places to perform experiments publicly and as part of a collective.[77] The Royal Prussian Academy of Sciences (a.k.a. the Berlin Academy) was supposed to be one venue—and it plays an important role in the story of Halle's Orphanage. Still, most of its membership participated through correspondence, making it difficult to sustain a productive forum.[78] The chambers (*Kammern*) that managed the natural resources of territorial leaders, testing new mining technologies or energy sources, were known as "secret spheres."[79] *Adel* or *Ritter* academies offered elite young men access to technologies and instruments, but the sons of the middling or professional orders—doctors, lawyers, teachers, preachers, engineers, craftsmen—needed access to such facilities as well.[80] Halle's Orphanage included schools for them all, as its administrators aimed to make this knowledge accessible to many more people. It even offered guided tours.[81] Because of its close relationship with the University of Halle, the Berlin Academy of Sciences, and the Prussian court, the Orphanage was able to contribute in substantive ways to ongoing efforts to change the way people thought about education, research, and public knowledge.[82]

Endeavoring to make information and new technologies public fit quite well with the modernizing strategies then being pursued by territorial leaders. Indeed, promoting new kinds of scientific sociability in Prussia was closely connected to the university reform initiatives mentioned above. Since Francke opposed scholastic methods, he was an ideal candidate to direct such an endeavor. It also makes sense that his organization would promote "eclecticism" as a way of helping others see the benefit of tolerating other perspectives so that productive debate and discussion would be possible. A revival of interest in eclecticism in Halle is key to understanding what set the Orphanage apart as a scientific community. It informed efforts to transform young people's understanding of what sciences were most useful and how they should be taught.

Eclecticism as a method of election or choice in philosophy has generally not received much attention from historians of early modern science.[83] This is largely because there was no standard methodology. Orphanage administrators were trying to create one, but in the end it was basically left up to the individual. In Halle describing oneself as an eclectic was a way of referring to the related pursuit of experimental physics and practical mathematics.[84] Like the word *Pietist*, the term *eclectic* was contested, its meanings in flux—and this actually became a problem in Halle, leading Wolff to express concern about it a few years before his expulsion from the city. I will go into more detail about eclecticism later; for now, I want simply to stress that eclecticism was broadly conceived of as a tool for assimilating perspectives and observations and a way of improving one's ability to become a discriminating observer.

Because of the link between eclecticism and observation, those teaching and training in Halle's Orphanage contributed to the transformation of scientific observation.[85] Increasingly, observation was becoming a "way of life," hardly something that almost anyone could undertake without a great deal of preparation and determination.[86] "When observation was reconceived in early modern Europe as the province of doctors, scholars, naturalists and other literate elites," writes Lorraine Daston, its status as coordinated, collective activity changed dramatically: "authored observations were systematically gathered by individuals, governments, mercantile corporations and scientific societies."[87] Orphanage administrators were aspiring to build a scientific society and, ultimately, a long-distance corporation—much like the Catholic Society of Jesus; this meant taking observation very seriously as a tool that could help their trainees negotiate a growing economy of scientific knowledge.[88]

Increasingly standardized observational procedures included repetition—observing the same phenomenon over and over again with a variety of precision instruments—note taking, synthesis, and description of results.[89] These activities required carefully cultivated ways of seeing (as well as hearing, smelling, and tasting) but also required patience, determination, and the disciplining of emotions. These activities also required increasingly specialized or technical forms of collecting. Observers from the Halle Orphanage collected cultural artifacts along with descriptions of plants, animals, and minerals in the languages of the places they traveled—especially Russia, India, and North America. They used names, local histories, and accounts of peoples as tools for understanding the relationships between place and species or kinds considered indigenous to the region.[90] They helped to generate acceptance of scholarly activities requiring new forms of instrumentation and technological expertise.[91] In sum, Halle's Orphanage was able to fill a growing need for venues in which to train *Observators* (to borrow Leibniz's term): young men with the expertise in mathematics, natural history, and physics needed to observe skillfully at home and abroad.

———

So much was going on inside the Orphanage between around 1700 and 1730 that it is impossible to explore every aspect of it in a single study. In the pages that follow, I have selected components I believe to be central to elucidating its importance as a scientific community.

In chapter 2, I explore conversations between Leibniz, Tschirnhaus, and Francke about how best to teach mathematics and experimental physics in the Orphanage. Tschirnhaus in particular spent several weeks in Halle, where he collaborated with Francke, student teachers, and administrators just as the Orphanage was getting off the ground. Tschirnhaus, who was a good friend of Baruch Spinoza, a member of the French Academy of Sciences and perhaps the most famous mathematician alive at the time, also considered himself a follower of Spener and was very interested in contributing to the design of the Orphanage's curriculum. Leibniz also collaborated with Francke regarding how to use the combined resources of the Orphanage and the Berlin Academy, which Francke joined in 1701, to attract the attention of Tsar Peter the Great. All of these individuals shared an interest in how best to train skilled scientific observers—traveling missionaries, chaplains, and teachers—who could apply their expertise while gathering information and taking measurements abroad.

In chapter 3, my focus is on how and why the Orphanage became implicated in the politics of confessional union that in many ways defined Brandenburg-Prussia after the Thirty Years' War. Efforts to devise institutional strategies for eliminating confessional differences led administrators to the writings of Johann Amos Comenius and to his system of pious natural philosophy, or "mosaic physics." The chapter concludes with a discussion of debates that emerged in Halle on the eve of Wolff's expulsion surrounding how to practice eclecticism. These debates, combined with Wolff's growing boldness, led many to embrace "exclusionary eclecticism": a form of eclecticism that involved strategically rejecting some perspectives and embracing others when creating theories and synthetic frameworks, rather than finding ways of respecting all points of view.

Teachers in Halle's model community were particularly excited about the potential of material models. As material ideas and agents of awakening, models were supposed to facilitate an intuitive way of understanding and to trigger desires to do certain things. In 1716, a local clergyman-mathematician and mechanical philosopher named Christoph Semler built one of the largest wooden models of Solomon's Temple ever exhibited in this period. The original was put on display at the Orphanage, and a copy of it was used during observational exercises in the *Pädagogium*. In the model, Semler combined and made visible the expertise of scholars from several confessional communities; he left a detailed account of the process he went through in order to build it. I consider this process in chapter 4 along with how it was used to generate ways of "conciliatory seeing" or "seeing all at once."[92]

In chapter 5 I tell the story of a highly skilled scientific observer, a theology student teacher named Christoph Eberhard, who was an expert in using the special *inclinatorium* that Francke showed to the king during his 1720 visit. This instrument caught the attention of both Tsar Peter I and the British Parliament, which had promised a sizable reward in exchange for a reliable method for measuring longitude. Eberhard's remarkable story provides us with a great opportunity to comprehend the kinds of activities young men associated with the Orphanage were engaged in and, more specifically, how they endeavored to use eclectic methods to assimilate information and to theorize about the observational data they had gathered. Eberhard's story also provides insight into how these efforts continued to be heavily politicized on the eve of Wolff's expulsion from the city.

To conclude, I explore efforts to copy the Orphanage in other contexts as a way of demonstrating the crucial role it played in generating scientific sociability, community, and expertise in central and eastern Europe. I look at several replica institutions built in the cities of Langendorf/Weißenfels, Sulechów, Zittau, Berlin, and Königsberg between roughly 1700 and 1750. By drawing attention to the institutional venues, practices, and objects through which admirers endeavored to copy Halle's Orphanage, my goal is to lay a foundation for further exploring their impact. Although each one of these replicas was unique, they remained linked together as part of a network in which the circulation of news reports and gifts played powerful roles. These institutions actively maintained ties to a scientific community of skilled observers.

TWO

Building a Scientific Community

Yes, this would be a kind of unrestricted orphanage where all poor orphans and foundlings could be fed, put to work and either raised to study or to participate in mechanics or commerce. GOTTFRIED WILHELM LEIBNIZ, "GRUNDRISS EINES BEDENKENS"

In early January 1698, one of Europe's most talented mathematicians traveled to Halle to see the Orphanage with his own eyes. He was a well-educated gentleman from Saxony named Ehrenfried Walther von Tschirnhaus. German scholars have long regarded him as a gifted natural philosopher who paved the way for the Enlightenment in the German states.[1] Yet today he is little known outside of Germany—and he may have wanted it that way. Tschirnhaus published relatively little, burned most of his personal papers before he died, and never held a formal post at a university. There is still an air of mystery surrounding his story and the nature of his interest and involvement with the Orphanage.

After leaving Saxony at the age of eighteen to study in Leiden, the young Tschirnhaus met Baruch Spinoza, who became a confidant and advisor. He traveled to England in 1674, where he met with Robert Boyle, Henry Oldenbourg, Isaac Newton, and John Wallis; despite their repeated requests for him to join the Royal Society, Tschirnhaus refused.[2] He joined the Académie des sciences in Paris in 1682 but never received a pension. Tschirnhaus probably became best known for his secret work on developing a recipe for porcelain that would result in Germany's first porcelain factory in Meissen, Saxony. A few scholars have stressed the importance of Tschirnhaus's writings to Christian Wolff, who tells a short but captivating story in his autobiography about his pilgrimage to meet the nobleman at the Leipzig book fair in 1705.[3]

These circumstances make Tschirnhaus's interest in the Orphanage especially puzzling. Why would this talented and highly sought-after natural philosopher, with contacts in virtually every European center of learning, drop by a provincial city on the fringes of Brandenburg-Prussia to help out with its schools? In this chapter, I set out to answer this question by considering Tschirnhaus's relationships with Francke and Leibniz, who very likely brought them together in the first place. In general, these three figures shared a mutual interest in Pietism—especially in collaborating with Philipp Jakob Spener—and in reimagining the school as an inspirational place of research.[4] Leibniz and Francke shared other interests too: both wanted to start a Protestant mission to China via Russia and saw the institutions they were helping get off the ground as necessary in order for this to happen. While Leibniz worked in Berlin to start a scientific academy whose members would direct the mission, Francke worked with Tschirnhaus to design programs that would introduce missionaries in training to mathematics and physics. Because of his efforts to help Leibniz achieve his goals, Francke became one of the earliest members of the Berlin Academy of Sciences; he welcomed Tschirnhaus to Halle because the nobleman had the expertise he needed to turn the Orphanage into a scientific community.

The New Way

Tschirnhaus and Francke spent roughly two weeks together in early January of 1698, but very little is known about exactly what transpired during their meetings.[5] Letters they exchanged afterward provide some insight into the content of their conversations. On January 17, 1698, Tschirnhaus wrote to say that he was pleased with Francke's plans to generate moral improvement through the reform of education. He explained that he was especially excited about efforts to include lessons in mathematics and other "good sciences." In his view, this was a big part of what made his aspirations so unique, since the study of mathematics was usually neglected in most secondary schools (*Gymnasia*); "Euclid and Archimedes are currently as good as banned" from these venues, he noted, and the "consequences were unbelievably devastating."[6]

Tschirnhaus claimed to have developed a "new method" for teaching mathematics, which he described in a letter of January 17. The method, he insisted, involved choosing to follow a new path or to go "the other way." As he explained:

There are particularly two ways of acquiring knowledge in the world: one way may take ten years where one learns a lot, but this way is not worth much in respect to the other way. The other way is cast aside by most but it is the best. One learns the genuine method of demonstration and how to distinguish the true from the false through practical experience. In a way that does not happen through the teaching of logic, one acquires the right method for finding the truth inside oneself—including how to meditate well, to sit still and to be attentive (strange qualities for youth).[7]

Going "the other way" involved acquiring proficiency in a "genuine method of demonstration" wherein one learned how to discern universal truth by "going inside oneself." Tschirnhaus's method, in other words, was inextricably linked to meditation and other forms of spiritual exercise perfectly suited to the aims of those involved in founding the Orphanage.

In this same letter, Tschirnhaus indicated to Francke that he had written him a summary of his method. He called it a *Basic Guide to the Useful Sciences, Especially Mathematics and Physics* and indicated that this was a special kind of gesture, indeed a "sign of a special kind of friendship and trust."[8] The *Basic Guide* was essentially a shorthand version of Tschirnhaus's *Medicina Mentis* (Medicine for the mind), first published in Amsterdam in 1687.[9] The *Medicina* was an expression of his interest in helping to develop a new science of the mind, in keeping with the efforts of his friend and mentor Spinoza. As Stephanie Buchenau has explained, it provided an opportunity for the nobleman to tinker with conventional explanations for the causes of genius, the tools or instruments of reason, and the meaning of the "art of invention," which he defined as a "general science by means of which whoever possesses it can discover not only everything hidden in mathematics but even everything that is hidden in the domain of understanding."[10]

In the *Medicina*, Tschirnhaus insisted on the importance of a middle path between a posteriori and a priori reasoning through a process of inner "recognition"; he argued this process "provide[d] access to an objective knowledge of ideas, different from the knowledge concerning the object of the idea."[11] Tschirnhaus's goal was to find a way of reconciling a growing rift between those who preferred empirical methods and those who adopted a rationalist approach to knowing the world. And, as Buchenau notes, this meant he was willing to concede (in opposition to Spinoza) that it was possible to deploy one's instruments of "sacred reason" without a "distinct knowledge of their nature."[12] A key implication of his claim was that "unmethodical empiricists and artisans are capable of acquiring true notions of things, and of developing their cognitive instruments, without necessarily being aware that they are doing so."[13]

Because of his ties to Spinoza, who was often described by contemporaries as an atheist, Tschirnhaus was widely criticized for the *Medicina Mentis*.[14] However, this did not prevent Spener from defending the mathematician and reaching out to him directly to share his thoughts about the treatise.[15] Spener even encouraged his son, Johann Jakob, to go study with Tschirnhaus because of the kind of training in mathematics he believed he was able to offer.[16] The nobleman also conveyed a genuine interest in Spener, particularly his writings and reform efforts. In the *Basic Guide to the Useful Sciences*, Tschirnhaus emphasized his interest in helping individuals learn how to cultivate an instinct or drive that would enable them to apprehend divine messages.[17] Because most people "so often only pay attention to outward appearances," Tschirnhaus wrote, "they miss the feeling of godly operations and often even find themselves to be diametrically opposed to them, so in the end nothing happens."[18] As a remedy he recommended reading the Bible and the writings of people with spiritual gifts—especially the writings of Spener.[19]

Francke was Spener's protégé and in some ways his closest friend and ally. By 1685, Francke had earned his *Magister* in Hebrew grammar and philology at the University of Leipzig, where he also founded a *Collegium philobiblicum*, focused on biblical exegesis.[20] He met Spener shortly after this and embarked on a drawn-out conversion experience, which he described as a "hard-won struggle."[21] After spending some time away from Leipzig and some quality time with Spener in Dresden, he returned in 1689, where he began to offer lectures at the university in German, not Latin.[22] His lectures, which were extremely popular, unabashedly criticized the way in which theology was being taught, including the lingering culture of disputation and emphasis on scholastic philosophy. That same year, several orthodox Lutheran theology professors, who were offended by the things Francke had said in public and saw him as promoting disorderly forms of behavior, formally investigated him (and his friends), banned the student conventicles, and refused to allow him to teach.[23] Francke left Leipzig in 1690. He took up a pastorate in Erfurt for a year, got in trouble for being too radical there too, and then, with Spener's help, received a pastorate in Glaucha, a suburb of Halle, in 1691.

Around the time he first met Francke, Spener took up a coveted post as senior court chaplain to the Saxon elector, Johann Georg III, in Dresden.[24] However, he did not stay long because of his constant clashes with the

Saxon elector over what Spener considered an excessively secular court culture. After this appointment, he was invited to serve as the Hohenzollerns' court chaplain in Berlin, where he learned of Friedrich I's plans to found a new university in Halle. Spener saw to it that Francke would also be offered a chair in Greek and Oriental languages when the new university opened in 1694. It was upon commencing his duties in Halle that Francke founded the Orphanage and opened his doors to Tschirnhaus.

After their first meeting and early exchanges, close associates of Francke, namely Johann Julius Elers and Baron Carl Hildebrand von Canstein, also began to communicate with Tschirnhaus regarding elaborate plans to manufacture glass in the Orphanage.[25] They wanted to make it easy and convenient for young people to make their own lenses for optical instruments; they especially wanted to know more about Tschirnhaus's experiments with the production of porcelain.[26] Elers made at least one trip to Dresden to learn from Tschirnhaus about how the process worked. In his remarkable "Journal of the Trip to Dresden" from 1704, Elers recorded very specific descriptions of what he observed. He spent hours with Tschirnhaus in glass-making huts (*Glashütten*) exploring "new strategies for making the fire grow hotter" and how to make different kinds of ovens.[27]

Relations between the nobleman and the Orphanage began to deteriorate when Tschirnhaus felt increasingly pressured to reveal his secret for making porcelain—he never did.[28] It did not help that Francke and his associates stepped up their efforts to purchase these secrets at a moment when Tschirnhaus and his estate (Krieslingswalde) were falling into debt.[29] Despite these complications, Orphanage teachers took his pedagogical advice quite seriously. A variety of reports about the curriculum stressed the importance of following Tschirnhaus's suggestions or, more specifically, applying his method.[30] In a 1702 description of what was being taught in the *Pädagogium*, for example, Francke wrote: "We are trying (in mathematics) to develop the Method that has been published by one well versed in this knowledge and a certain very famous statesman who knows our *Pädagogium*."[31]

After he wrote his *Basic Guide to the Useful Sciences* for Francke, Tschirnhaus sent a copy of it to Leibniz. From Berlin, Spener also tried to ensure a copy landed in Leibniz's hands. Leibniz wrote to Spener with gratitude on June 8, 1700, from Berlin but noted that Tschirnhaus had also sent him a copy of the text.[32] He said he wished Tschirnhaus had gone into more detail but overall felt the text contained some "good and useful" prescriptions. Leibniz went on to say that he was especially glad to learn of Tschirnhaus's efforts to "build up the sciences toward virtue" and indicated he knew about Francke's efforts in Halle.[33] He then wrote a detailed review of the *Basic Guide* and sent a copy of it to Tschirnhaus.

In the review Leibniz described the short text as beautiful and offered his summary of its most useful insights. He stressed the importance of "taking the middle way" and using new pedagogical practices to help awaken and inspire young people:

Nowhere is the truth more apparent than in mathematics, which also can be usefully applied to water and land in war and in peace, but even more because mathematics transmits before the eyes key proofs of the pure truth and shows the way to correctly recognize this truth. . . .

. . . Two ways of practicing mathematics are currently followed: one way uses many instruments, praxis and books but does not bring one closer to the truth. The other way uses the theories of Euclid, Archimedes, Apollonius and Analysin, which seem at first to slow up things because of the emphasis on major findings from the past, but only this way provides the grounds for one to find oneself—something that others search for with many costs. Indeed, because of what the youth want, *one must take the middle way* and teach the beginners only the most important practices from the discipline of mathematics in order *to awaken a passion and a light within them*, after which they can be shown causes and led through theory.[34]

Leibniz got right to the point about Tschirnhaus's method: it marked a compromise between two ways of practicing mathematics. One way started with "instruments, praxis and books." The other way was more theoretical—but very time consuming. The middle way was a new way that was supposed to combine both of these approaches. It was also supposed to be fun. In the final portion of his review, Leibniz echoed Tschirnhaus's sentiment that "if one were to teach these things in the schools, learning would become *recht Ludus*," or real play.[35]

Tschirnhaus saw a conflict between the need to teach young people by making them memorize things—generally not fun—and the need to inspire them with real objects and instruments. One was primarily an intellectual enterprise involving a lot of rote learning. The other was more imaginative and awe inspiring. How to combine them? In pursuing this problem, he took issue with some of the boldest claims of Spinoza. Tschirnhaus did not accept his friend's strict separation of the intellect from the imagination. This careful separation in Spinoza's philosophy lay at the heart of his criticism of Boyle's experimental philosophy, for example, which Spinoza felt relied much too heavily on the faculty in humans most prone to producing errors and "did not put forward any mathematical proofs." Steven Shapin has explained that "there was, in

Spinoza's view, a fundamental contrast between experiment and mathematics in their respective abilities to secure conviction"; because Boyle did not offer mathematical proofs of his experiments, they could not convince. Boyle countered that he was right to be wary of mathematical proofs and to "consult sense" in order to check "whether men have not been mistaken in their hypotheses and reasonings."[36]

Tschirnhaus claimed to be offering an alternative to these two extremes.[37] He advocated "the use of a good method in which the intellect, imagination and senses cooperate," which he said "would lead to the ideal situation where the school can become *ludus*."[38] He believed schools would then become sites of inspiration, where awakened children would find themselves compelled both to use the instruments they observed in demonstrations and to acquire as much advanced theoretical knowledge as possible about the three essences (*Wesenheiten*) he had identified as always at work in the world.

In the *Medicina Mentis*, Tschirnhaus described these three essences as *Imaginative/Imaginables*, *Mathematical/Rationales*, and *Physical/Reales*. The following is a brief summary of how he understood the distinctions between them:

When I think about everything that I have become acquainted with since childhood through seeing, reading or hearing, and I reflect awhile on the order in which I acquired these thoughts, a kind of chaos emerges. . . . Some of these thoughts are strong essences, which, it appears, are taken as truth by me earlier than they are understood. These often influence my spirit . . . as appearances that act upon me from the outside . . . in active ways. . . . I will name such appearances that act upon me in this way "Imaginables" or, if one prefers, *Phantasmen*.

Some things I grasp in different ways, for example, what I know from figures, numbers, movements and similar things. . . . Such essences, which . . . appear to have no existence outside of my person and of which I grasp only the clean or very abstract . . . I will call "Rationales" or "Mathematical."

Finally, I observe that I have thoughts from some essences, which, unlike the preceding Rationales, can be grasped in multiple ways rather than in singular and unchangeable ways. . . . For example, all that we grasp as material, unclean or nontransparent. These essences I will name "Reales" or "Physisches."[39]

Different techniques or methods were required to better understand *reales*, *rationales*, and *imaginables*. Mathematics provided opportunities to understand *rationales*, just as object lessons afforded new insight into *reales*. Tschirnhaus optimistically believed that the key to understanding *imaginables* was through experimentation.

Tschirnhaus explained in his *Basic Guide* that one of his most basic concerns in the *Medicina* had been to "expose the obstacles and what comprise the largest and most comprehensive errors" in the act of "recognizing Truth," both "through the senses and through the power of the intellect."[40] To alleviate the tendency of the senses to produce errors, Tschirnhaus offered his readers a set of experiments they could perform in order to better understand their *imaginables*. "One learns about imaginable essences by doing countless careful experiments," he wrote, "which should be performed according to specific rules."[41] He advised his readers to consider the rooms in which they were going to observe their *imaginables*, "in part so that we can see the nature of the outer or surrounding body and its influence upon another . . . and in part so that we recognize in which order the experiments, or attempts, should be made."[42] Designating spaces in which to investigate *imaginables* was also important to Tschirnhaus because "when someone sets aside a room for the teaching of some thing, that thing is more quickly learned."[43]

Those who performed experiments in an "empirical manner," like Robert Boyle, he wrote, were so concerned with understanding the properties of some real thing that they did not use the site of the experiment to consider the *imaginables*. They missed an opportunity to reflect on the nature of perception and the workings of the imagination. For example, as he had explained in a much earlier exchange with Leibniz, the microscope not only represented an opportunity to discover more about the smallest parts of large things; it also offered new ways of observing the thing in its entirety. One could explore the effects that different kinds of light had on one's perception of what one saw while using the instrument.[44] The microscope was more than a tool for extending or enhancing vision; it was a tool with which to study how we observe and perceive the world—and for teaching others how to observe in highly skilled and nuanced ways.

In the *Basic Guide*, Tschirnhaus identified four steps he hoped Francke would implement in the schools of the Orphanage and *Pädagogium*. First, and "above all," Tschirnhaus wrote, "one must teach the beginners the most important practices from all of the mathematical disciplines." To "awaken a love of mathematics," the teacher must appeal to children's tendency to accept information revealed to them by their senses. Children were to be taken outside to measure fields or taught how to make their own sundials and optical lenses.[45] The full implication of Cicero's insistence that the context of discovery was more important than the context of justification became clear in Tschirnhaus's first step. Here was the engine of invention and, as Cicero recognized, the beginning of practices necessary to heal the mind.[46] The next two steps involved using the mind to cultivate an ignited passion for mathematics by moving into more the-oretical terrain. By the fourth and final step, one would be ready to begin an intensive study of physics through experimentation.[47]

A Collection of Curious Things

When he wrote his appraisal of Tschirnhaus's new method around 1700, Leibniz signaled his approval of a previous generation of educational re-formers who had argued persuasively that schools needed to be more fun and provide young people with access to useful information.[48] In the mid-seventeenth century, Leibniz's former teacher in Jena, a mathematics pro-fessor named Erhard Weigel, advocated founding special kinds of schools for instilling virtue in children through the mechanical arts by showing them *naturalia*, machines, measuring instruments, and models.[49] The Czech educational reformer Johann Amos Comenius, whom I discuss in the following chapter, was also an advocate of reimagining the school as a repository of curious objects and information from which children could learn more actively and directly about the world.

As he read Tschirnhaus's *Basic Guide*, Leibniz's plans to found a scien-tific society in Berlin were gaining momentum.[50] He had been working toward this goal for some time and produced several schemes in which he worked out exactly what he wanted this society to accomplish. In a very early "sketch of a thought about establishing a new society" (1671), Leibniz played around with ideas that were similar to those expressed in Thomas More's *Utopia*, Tommaso Campanella's *City of the Sun* (*Civitas So-lis*), and Francis Bacon's *New Atlantis*.[51] In a stream-of-consciousness style, he described a new society that aimed

to improve the schools, to exercise the youth, to teach youth languages and the real-ity of the sciences to prevent difficulties when traveling, to develop a kind of gentry school institution, to make handiwork easier with certain advantages and instruments, with constant and inexpensive fire and motion try and make everything in chemistry and mechanics, and in glassmaking, perspective, machines, water arts, clocks, turn-ing machines, painting, book printing, dying, weaving, steel and iron works . . .

. . . to acquire foundations and collections of curious things, a theater of nature and art or an art- and rarity- and anatomy-chamber for the easy learning of all things, to form new apothecaries, gardens and libraries, . . .

. . . to collect lost relations, experiments, and letters of correspondence, to bring everything in order, provide poor students with room and board, to create an insti-tution where the work of these students will be of use to them and to the society[52]

Here we see Leibniz describing a kind of scientific society that bears a close resemblance to the new "Foundation" described in Bacon's *New Atlantis*—also known as Solomon's House.[53] Yet in Leibniz's schematic musing, this foundation was an elite scientific society that contained schools: it was committed to wide-ranging educational reform, including a new vision of the school—separate from the university—as a repository, a "theater of art and nature," and a place for research.[54] University students were to work as the teachers and receive room and board in exchange for their help. The Foundation was to have an apothecary, botanical gardens, a library, an archive, collections of curious things, and instruments that would not be off-limits to young people.

Roughly thirty years after he wrote it, Leibniz's "sketch of the thought" was much more than a thought: it was in the process of being realized in Berlin and in Halle. Indeed, soon after its founding, the Orphanage became a kind of repository for all manner of natural and artificial things—and this is how contemporaries described it.[55] By 1720, the Orphanage had its own apothecary and chemical laboratory and a library and archive for collecting "lost relations, experiments and letters of correspondence," and its collections of "curious things" had expanded dramatically—exactly as Leibniz had hoped. The Berlin Academy did not have these things; in fact, as noted in chapter 1, members participated through correspondence and rarely had the opportunity to study things or conduct experiments together.

By the end of the seventeenth century, possessing this kind of collection, a curiosity cabinet, or *Kunst und Naturalienkammer*, was a way of signaling an organization or individual's high status and worldliness, including an interest in participating in conversations about natural history, place, improvement, utility and universal knowledge. It also signaled an interest in efforts to devise new frameworks or mechanisms for making sense of this diversity.[56] Possession of such a cabinet meant that one had access to a container of raw evidence necessary in order to deploy systematically a range of observational and experimental procedures—perhaps using the instruments that were also on display. Containing or capturing objects from near and far marked a way of investing diverted commodities with a "new cognitive significance," marking them as places to direct the attention and to use in a variety of tests.[57]

In the case of the Orphanage, these objects were set aside for use in lessons; for example, a description of the *Pädagogium*'s curriculum from 1702 notes that every week objects from the organization's collections were shown and explained to pupils "according to their nature, qualities, utility and applications."[58] Teachers also introduced their pupils to other kinds of natural objects, such as "things found in an Apothecary generally not known to everyone," and watched while a variety of experiments were performed with the air pump in particular.[59] In the early years, Orphanage administrators acquired approximately 150 objects, many of which friends and benefactors had sent as gifts.[60] In several instances, they grew the collection by accepting rare or unusual objects from guardians looking to secure their children's admittance to the Orphanage schools.[61] Several objects, including a unicorn's horn, the horn of a rhinoceros, and three whale teeth, were sent directly from Berlin as gifts.[62]

As the Orphanage expanded, it was able to send out growing numbers of missionaries, tutors, teachers, and household pastors. Predictably, the flow of objects into the Orphanage's repository grew rapidly, so that the museum contained well over five thousand items by the early 1730s. The collection became so unwieldy that administrators moved it to the top floor of the main building, where it can still be viewed—relatively intact—today. Among the things one might have seen during a visit were conch, snail, and mussel shells; a variety of insects, salamanders, and chameleons; a scorpion; a tarantula; a three-and-a-half-foot-long worm that a medical student pulled out of the foot of a patient; a tumor from the chest cavity of an eighteen-year-old boy; the stomach and intestines of people who had died from various of illnesses; an assortment of body parts; a skeleton of a baker's wife who had admitted under torture to her involvement in a plot to murder a rich widow and had agreed to donate her body for dissection after her execution; a crocodile (14½ feet long); the rib of a whale (18 feet long); a cabinet full of minerals and precious stones found in mines in Hungary, Saxony, India, and Sweden; over 120 types of marble sent from Venice; and a four-foot-long leaf from a coconut tree.[63] The collection also contained a variety of cultural or ethnographic objects, including a Chinese patent and a letter from Martin Luther to the king of Denmark, clothes, and other artifacts (knives, spoons, jars) from Greenland, Russia, China, and Turkey. Objects sent from India included European edificatory books that had been translated into Tamil, texts written in Tamil on palm leaves and the writing implements used to make these inscriptions, a "Malabarian fly swatter made of peacock feathers," a "red idol-box, painted with many figures of gods," and a pair of "penitence slippers" made out of wood and nails.[64]

Efforts to expand the collection of cultural artifacts were inextricably linked to Orphanage administrators' own efforts to extend their global reach—in effect to become the kind of "seminar of nations" Francke described in one of his earliest descriptions of the Orphanage. Francke and Leibniz also aspired to create a Protestant missionary society, whose affiliates would have the same expertise associated with Jesuit "scientific missions" abroad.[65]

Leibniz, Francke, and the Berlin Academy's Mission

Leibniz's plans for the Berlin Academy and Francke's vision for the Halle Orphanage involved establishing a Protestant mission. The two entered into a conversation about this common goal, exchanging at least eight letters between 1697 and 1699.[66] The conversation started when Francke sent his friend Heinrich Neubauer to the Netherlands to observe and report on different institutional models that might be of use to those building the Orphanage.[67] On his way to Holland, Neubauer stopped in Hannover, where he brought Leibniz a copy of Francke's *Historic News about the Care of Poor and the Rearing of the Youth,* one of his many attempts to describe what he was up to in Halle.[68] During his meeting with Leibniz in Hannover, Neubauer also relayed Francke's approval of Leibniz's *Novissima Sinica,* in which Leibniz had also presented a detailed plan for a Protestant missionary enterprise to China via Russia. The *Novissima Sinica* contained letters and updates from Jesuit missionaries about the Chinese "Tolerance Edict" (1692), the Russian-Chinese Peace of Nerčinsk (1689), and the Chinese emperor's interest in astronomy. In it Leibniz also outlined a project of cultural reconciliation that made Russia, specifically Tsar Peter the Great, the critical bridge.[69] Francke—via Neubauer—wanted Leibniz to know that "the field in Russia had already been opened" and that the Halle Orphanage was strategically necessary for accomplishing the goals Leibniz had in mind.[70]

Leibniz responded with a long letter to Francke in which he made clear that he fully appreciated the powerful points of convergence between their plans. He said that he saw the Orphanage as the culmination of several attempts by others (he specifically mentions Erhard Weigel) to teach useful sciences and expressed his hope that several large cities would also adopt the model Francke was introducing in Halle.[71] In this same letter, Leibniz recommended that Neubauer go to Hamburg to meet with Leibniz's correspondent there, Vincent Placcius, so that the two could discuss other issues related to reforming Germany's schools; there, he suggested, a propaganda campaign should be launched promoting the founding of new schools after the "Halle model." He also indicated that he immediately recognized in Francke's plans a set of educational objectives already being pursued in similar projects under way in Paris and "all over the place." These were private academies, organized by "skilled men and pensioners"—sometimes called upon to do so by the state—who expressly modeled their plans after Jesuit colleges.[72] Leibniz situated the new project in the pan-European context he knew well, but he also made a point of indicating how the Halle model was different from those other academies "in Paris and elsewhere." Leibniz was especially pleased, he wrote, that Francke would be reaching "young men before the age of thirty" with "the new method," because wherever this method was applied, "cognitions are turned into expressions that cannot be called into doubt."[73]

In the preface of the *Novissima Sinica*, Leibniz argued that both China and Europe could improve themselves if they became open to what they had to teach one another. He also made extensive use of Plato's link between virtue and geometry, where, as Patrick Riley explains, "knowledge of geometry (as something 'amounting' to a faith) is crucial."[74] Teaching young people higher mathematics, in other words, was a form of "spiritual exercise"—and ultimately a way of generating the forms of spiritual awakening and improvement that Spener, Francke, and others in Halle were seeking.[75] In the *Novissima* Leibniz also offered a highly accessible description of philanthropy.[76] He described the Chinese people as model philanthropists, who had managed to effectively realize wisdom in their daily lives. They "behave to each other so lovingly . . . almost in emulation of the teachings of Christ," and are "averse to war."[77]

Francke went out of his way to make sure Leibniz knew he approved of the *Novissima*—including these sentiments about Chinese morality and philanthropy. This is important to keep in mind, considering that Leibniz's protégé, Christian Wolff, would be roundly criticized for making similar kinds of statements about Confucian morals in his famous 1721 speech.[78] In the early years of the Orphanage, though, Leibniz and Francke saw themselves as allies who had much to gain by joining forces. In fact, their remaining correspondence concerned how to strategically maneuver to make contact with the individuals closest to Tsar Peter the Great.[79] The tsar made his first tour of Europe in 1697, and both Leibniz and Francke saw the visit as their chance to secure the tsar's personal support of their mission plans. Upon finding out that the children of the tsar's general had been placed at a school in Berlin while their father was touring, Leibniz asked Francke to recommend to Spener that he look in on them occasionally.[80] As court chaplain for the Hohenzollerns, Spener would have been in a position to extend them special privileges. Leibniz and Francke also attempted to convince Tsar Peter to come to Halle so that he could see the Orphanage with his own eyes.[81]

Leibniz invited Francke to join the Berlin Academy after it was officially inaugurated in 1701. That same year he invited another mutual friend, a theologian working as a court chaplain in the Prussian city of Königsberg, Conrad Mel, to join as well.[82] Mel corresponded with Francke about his plans to found an orphanage in Hersfeld modeled after Halle's, which opened in 1709.[83] As late as 1711, Mel had also sent Francke a copy of a catalog containing descriptions of *naturalia* and other curiosities from his own collection, which he was attempting to sell in order to raise funds for his own institution.[84] Prior to this he had made his support of the idea of a Protestant mission well known, producing his own rationale for linking such a mission to ongoing efforts to start a new scientific society in Berlin. He explained this rationale in a text he called the *Point of Observation of the Evangelical Legation*. In it he explained that the purpose of the Berlin Academy included not only the study of nature and all manner of sciences but also sacred missions.[85] He also offered a detailed plan for preparing a new mission that included teaching mathematics in academy-sponsored training institutes. These institutes were necessary, he argued, because it was clear that Jesuit missionaries' expertise in mathematics had been the key to their success in China:

The mathematical sciences deserve to be paid attention to above all else because the kind of progress the Roman Catholics have had through teaching them is well known. It is to be advised that one should look for the best mathematicians from all the universities, who can present their thorough understanding of these arts so that they can (unlike before) at least challenge the Roman Catholics' superiority.[86]

Mel stressed that despite their success, Jesuits used their knowledge of mathematics to perpetuate divisiveness, not friendship, which was why their efforts needed to be challenged:

Experience has consistently shown what the Jesuits have been able to accomplish through their mathematical arts. Through their knowledge of the arts of observing the stars they were able to put their foot in the middle of the Chinese kingdom; . . . but In discovering new knowledge of the entire earth, they have furthered a divided Christian Israel among the heathen Canaanites.[87]

Even though they would be relying largely on their example and taking cues from their success, the Protestant mission Mel wanted to be associated with the Berlin Academy was supposed to help promote confessional union. His sentiments reflect his own engagement with state-sponsored efforts in Prussia to eliminate confessional difference (see chapter 3).

Young men who attended the special schools Mel described were to receive training in a variety of languages, including Russian, Chinese, and

Arabic—all languages to be offered in the Orphanage.[88] Wherever possible, they would also learn directly from academy members, who would be mainly professors and gifted university students:

Preparation of an evangelical mission involves founding an Academy or School where the persons destined to become missionaries can be trained.

I. The persons who will direct this large work are:
1. Our king. . . .
2. The Brandenburg Society because the members are sought after in all sciences for their learning and have a basic knowledge of things abroad. . . .
II. The teachers of this Academy will be
1. Exceptional professors from different locations, faculties and sciences.
2. . . . Exceptional Students from Academies and from all kinds of faculties and sciences, who are recommended by the Professors, examined and approved by the Society, . . . and who have a noticeable fear of God.[89]

Sound familiar? It is not too great a leap to suggest that the Orphanage schools and the institutes Mel describes here were actually one and the same.

Indeed, the Orphanage, the University of Halle, and the Berlin Academy were all part of the same package, designed to do different things but to work together toward the same goals. With the exception of the university theology faculty's relationship with the Orphanage—a relationship that became more pronounced and carefully controlled over time—these three institutions eventually grew more self-sufficient and disconnected from each other. But these early efforts to make a robust training program in mathematics and physics a central part of the Orphanage's identity are crucial to understanding how it came to function as a scientific community. The program and pedagogies that both Leibniz and Tschirnhaus recommended stayed in place long after they stopped communicating frequently with Francke and his administrators.

I know of only three letters exchanged between Leibniz and Francke after 1699, all in 1714. They seem to indicate that more conversations transpired between the two men—although it is difficult to know for certain if or how frequently they remained in touch.[90] Read with evidence of the important connections between the Berlin Academy, the university, and the Orphanage in mind, these letters speak to the pressing question of what the word *Pietism* meant to both Leibniz and Francke as they attempted to define it for each other.

Francke, writing from Halle on January 10, 1714, started their last exchange with a bold request for Leibniz's help with salvaging the word *Pietism*, which, he noted, had been tarnished unjustly:

Now, your Excellency knows for himself, so that a demonstration will not be required, how "Pietism" is nothing other than a name for men who are obliged to respond to morally unsound men through the study of the literal word and truth of piety, thereby restoring and increasing credibility. . . . This name first came into being in Leipzig (and the Leipzig theologians have provided their own account of it, I have attached one of their notices); there around 1700 a Commission was established to investigate the entire thing . . . something was published . . . that attested to the purity of our theological readings but Pietism was still publicly declared to be false and slanderous. . . . Your Excellency will not only be familiar with the holy D. Spener's *Theological Concerns* and his edited history of Pietism, but also the famous tract by H. von Seckendorf . . . both of which place the entire affair in a clearer light.[91]

Francke's specific request of Leibniz followed:

Now that your Excellency has been here in person and given me the honor of viewing my institutions for taking care of the poor and raising the youth, and has also spoken with me about them many times, and from this has undertaken a fundamental investigation of what people like to call Pietism, I would like to ask you to please, for the love of truth, to provide the king and his ministers with the proper picture of the entire saga.[92]

The letter is amazing for several reasons. In it, Francke explained that orthodox theologians in Leipzig had coined the term *Pietist* and used it in a derogatory manner. The letter also suggested that Leibniz and Francke were continuing to engage in conversations about the Orphanage and that they had done this in person. The request was specifically linked to Francke's discovery that two of his former students were being denied employment in the Kingdom of Hungary (Siebenbürgen, or Transylvania) because of their reputation as Pietists.[93] Leibniz was at the court of Karl VI in Vienna at the time, and Francke saw this as a great opportunity to lean on the emperor to intervene.[94] Indeed, the problem, as he and Leibniz both knew, was a much more complex one about the meaning of "what people like to call Pietism."

Leibniz had taken it upon himself, Francke noted, to undertake his own investigation of Pietism. In Leibniz's response—a clear statement of support for Francke indicating that he did ask the Austrian emperor to intervene—he defined Pietism as a set of practices with no underlying dogma:

It appears that people see your innocence but are seized by momentary delusions. . . . Because the Kaiser is interested in the circumstances in Siebenbürgen, I hope he will be able to right the wrong and that restitution will follow.

Thanks for sending me the Leipzig response against Pietists. How could one think that they are members of a new sect when they have no underlying dogma but instead focus on getting to know symbolic books and on practicing Christianity?[95]

Leibniz then made a very pointed statement about juxtaposing spiritual training with training in the new sciences:

In my estimation, the most important thing is that one strive to improve the education of the youth. And to do this it is to be hoped that the furthering of the sciences will be combined with promoting the fear of God and that this will shed more light on the way to a natural recognition of God.[96]

He linked these aims to sentiments he had expressed in his *Theodicée*, which he said a variety of people of different beliefs approved of. An example of someone who disapproved of the *Theodicée* was Johann Franz Buddeus, whom we will meet in the next chapter. Leibniz closed by mentioning that the Austrian emperor was "well disposed towards funding a society of science" and was seeking his advice in this regard.

In the letter that marks the end of this exchange, Francke conveyed his feelings about the *Theodicée* and voiced his support of Leibniz's sentiments. "I am of the opinion that furthering the fear of God and furthering the sciences are two tasks that want to be joined, indeed must be joined," he wrote,

and my few institutions, in which now 1700 children and scholars and 120 teachers can be found, show how this can be done. God bless your good advice, and that of the Emperor, for the funding of a society to honor his holy name and to bring about a unified church. As for the badge of truth that you have laid forth in your *Theodicée*, we rejoiced heartily to receive it.[97]

In his own estimation the word *Pietist* was really "nothing other than a name," Leibniz was a friend and ally, and the Orphanage was a key venue for the furthering of the sciences.

THREE

Negotiating the Irenical Turn

Leibniz, Tschirnhaus, and Francke engaged with a variety of issues addressed by a previous generation of educational reformers, including the Czech pedagogue Johann Amos Comenius. Inspired by Francis Bacon, Comenius put forward his vision for spiritual, moral, and social regeneration in several essays, including a collection of plays he wrote for young people to perform in schools entitled *Schola Ludus*, or *School as Play*.[1] A political exile for much of his life, Comenius traveled extensively throughout Europe, including to London, where he became part of the celebrated Hartlib circle. He was also a leader of a Protestant group called the Unity of Brethren (*Unitas Fratrum*) that promoted Christian irenicism, or confessional reconciliation.[2] As a way of realizing his irenical goals, Comenius developed his own system of pious natural philosophy—also referred to as "Christian philosophy" or "mosaic physics"—which he described in his 1633 treatise *Natural Philosophy Reformed by Divine Light*.[3] It hinged on the assertion that God had spread pieces of light throughout the world that needed to be reassembled "not through an encyclopedic systemization" but intuitively—through an "inner eye" of faith.[4] Comenius noted that he derived his physical system directly from the book Genesis.

Most who have considered Comenius's system of natural philosophy have concluded it was never realized anywhere—that he and those who followed him shared "an agenda rather than a practice."[5] Yet accounts of what was actually going on inside the Orphanage suggest otherwise. In addition to applying Tschirnhaus's methods for teaching mathematics and physics, Orphanage student teachers explored strategies for making a pious system of natural philosophy accessible as a set of practices, including teaching young people how to observe, gather information, and reassemble it responsibly. Bringing Comenius's program to bear on the Orphanage fit well with the irenical politics pursued by the Prussian state. Indeed, many contemporaries associated the Orphanage with the irenical turn: consider the epigraph to this book from Johann Ernst Bessler.[6]

Brandenburg-Prussia was one of the only territorial states in the Holy Roman Empire whose leaders supported institutional strategies for promoting confessional reconciliation after the Thirty Years' War. In this chapter I take a closer look at these efforts along with the related conversations about observing with the "inner eye" that they advocated. Spener and Francke were involved in these conversations, as were Leibniz and the cofounder of the Berlin Academy of Sciences, Daniel Ernst Jablonski, Comenius's grandson.[7] I argue that these efforts to synthesize perspectives or to reconcile confessions in Prussia went hand in hand with efforts both to revisit Comenius's natural philosophical program in the Orphanage and to teach eclecticism, a method of choice that empowered individuals to conduct their own investigations of nature and to make their own choices about what aspects of a particular point of view they found convincing.

The more serious administrators became about teaching Comenius's system in their schools, though, the more they had to face up to its biblico-historical literalism and to reevaluate the meaning of the term *eclecticism*. Did practitioners have the right to reject certain perspectives entirely? Or did they have an obligation to find the good in all perspectives? Was reason the primary tool of synthesis, or should moral principles and the "inner eye" direct this process? Largely because of the efforts of Jena theologian Johann Franz Buddeus, these questions persisted and ultimately led to a major curriculum revision.[8] Reflecting on this revision, which involved reimagining the importance of teaching mathematics (relative to history and ethics, for example) in the Orphanage, can help us better understand some of the issues at stake in the community on the eve of Wolff's expulsion.

From Toleration to Evangelical Union in Brandenburg-Prussia

The contributions of the theologian Georg Calixt to the confessional politics of Brandenburg-Prussia dramatically impacted the orientation of the discussions that began occurring in Berlin and Halle in the 1690s.

While holding a chair in "controversial theology" at the University of Helmstedt from 1614 to 1656—before, during, and after the Thirty Years' War—Calixt began to seriously weigh the pros and cons of promoting religious tolerance. The experience of this war had convinced him that state-mandated tolerance policies were useless and that more radical measures were needed to completely eradicate theological differences—and strife.[9] Forcing people to be tolerant of those of different beliefs did not automatically generate respect, he believed. It did not mean that competing groups would not react violently toward each other in certain circumstances—as the disasters of the Thirty Years' War had shown.

Calixt strongly criticized the famous *cuius regio, eius religio* principle of the Peace of Augsburg (1555), which was affirmed again in the Peace of Westphalia (1648).[10] While the peace allowed those who practiced a faith that was different from that of their region's established church to do so publicly during certain hours and in private at any time, the principle upheld a fundamental confessional hierarchy. It allowed a single set of spiritual practices associated with one community to dominate the other, and this, Calixt argued, was dangerous. He insisted that the *cuius regio, eius religio* principle should no longer be upheld and that vigorous interfaith discussions needed to take place that were oriented around devising a *consensus antiquitatis* based entirely on careful readings of the fundamental tenets of apostolic Christianity—readings performed by theologians, of course. Once everyone was in agreement about the common truths animating all of their various expressions of faith, he believed long-standing differences and resulting conflicts would fade away.[11]

Calixt's opponents, especially Abraham Calov, branded him a syncretist; they saw him as someone who wanted to do away with the unique characteristics of each faith.[12] Calov's criticisms led to a series of very heated disputes that pitted the theology faculties of the empire against one another.[13] The main concern was that by reverting entirely to a common source, syncretic methods undermined the integrity of varying practices, or the peculiar aesthetic that was the unique signature of each group. But the return to a true or apostolic Christianity, which also became one of the defining characteristics of Lutheran Pietism, was viewed in Calixtean circles as a serious and necessary step to the elimination of strife and restoration of what Calixt called a *Pax ecclesiae Germanicae*.

Of course, implementing Calixt's solution required more than careful readings of sacred ancient books. It involved inspiring all contemporaries, regardless of their social status, to monitor his or her practice of piety (*praxis pietatis*) so as to acquire "the true faith" and the message of love that is at the heart of the New Testament.[14] Devotional literature as a genre proliferated in the German lands at this time as contemporaries got caught up in this cause. Translations of English devotional literature sold especially well in German cities. One popular collection of exercises was Emanuel Sonthom's *Göldenes Kleinod*, which was a mostly copied version of the Jesuit Robert Parson's *Book of Christian Exercise*; other popular English-language texts included Lewis Bayly's *Praxis Pietatis* and Joseph Hall's *Arte of Divine Meditation*.[15]

Both Calixt and Calov were irenicists because they believed in the need for a union of Christian confessional systems. They were participating in what are sometimes referred to as the "syncreticist controversies"—still much discussed at the time of the Orphanage's founding. David Pareus, a reformed theologian active a generation before Calixt and Calov at the University of Heidelberg, then a center of Calvinist learning, outlined the concerns of irenicists with the publication of his *Irenicum sive de unione et synodo evangelicorum concilianda* in 1614. Scholars have noted that Pareus and the cause of confessional union must have inspired Comenius's program of pious natural philosophy.[16] Pareus also met the Calvinist theologian John Bergius, who became the primary religious advisor to the first three electors of Brandenburg-Prussia, befriended Calixt in 1645, and forged the first official policies of religious tolerance in Brandenburg-Prussia during and after the Thirty Years' War.

The electors of Brandenburg-Prussia, who were Calvinists ruling over Lutheran estates, became committed to further developing policies of toleration that actually functioned in the wake of this war.[17] In 1685, after the revocation of the Edict of Nantes, which had allowed French Calvinists, called Huguenots, to practice their faith without fear of persecution in France, Friedrich I issued the Edict of Potsdam, inviting those affected by the revocation to come to Brandenburg-Prussia. Approximately forty thousand Huguenots took him up on the offer and relocated to Berlin. Several rapidly acquired elite posts and fashioned themselves as Berlin's new literati. They joined salons and learned societies and circles. The most famous of these salons belonged to Ezechial Spannheim, who became the Hohenzollerns' court historian after Samuel von Pufendorf died in 1694. D. E. Jablonski attended the Spannheim gatherings often, as did other early Huguenot members of the Berlin Academy such as Etienne Chauvin, Pierre Dangicourt, and Johann Jakob Chuno.[18] Another highly prominent early member of the Berlin Academy, Maturin Veyssiere Lacroze, also frequently attended.[19] Several of these individuals were involved in the intensification of confessional union deliberations that occurred under Friedrich I shortly after their arrival, deliberations that influenced the orientation of the Orphanage's educational programming.

Brandenburg-Prussia's toleration policies were unlike those in any other German state. Yet because of the influence of Leibniz and other Calixt-inspired theologians at court, Friedrich I wanted to be even bolder. He became convinced of a need to move beyond existing policies of toleration toward confessional union. At the time, this was imagined as a very modern and forward-thinking thing to do. Leibniz had been working from his post in Hannover, in the court of Georg Ludwig of Brunswick-Lüneburg, to formulate new Calixtean schemes for confessional reconciliation since 1696 or earlier. A devoted irenicist, he was convinced of the need to unify all Protestants under a single confessional banner before pursuing reconciliation between Catholics and Protestants. He supported the founding of an evangelical church as a middle ground, a place where spiritual exercises and the liturgies they were based upon could be calibrated and formalized. This, he believed, would produce a unity of belief.[20]

Leibniz received a great deal of support in these efforts from his friend Jablonski, who had become Friedrich I's court chaplain in 1693. Jablonski served alongside Spener, who was also the senior rector of St. Nicholas's Church in Berlin and very involved in confessional politics. Six years after having taken on this post, Jablonski became a bishop of the Unity of Brethren; just like his grandfather Comenius, he was committed to its survival. Well aware of the legacy of Comenius, Jablonski championed the primary concern of the *Unitas Fratrum* (Christian union). In the 1680s, he spent two years on a brethren-sponsored scholarship in Oxford, where he became drawn to the Church of England. He was convinced that other than the Unity of Brethren, the Anglican Church's episcopate were the only ones to have legitimately succeeded the first apostles. The key to furthering the cause of union, for him, resided in strengthening ties between the *Unitas* and the Anglican Church.

As a court chaplain in Berlin, Jablonski was actually in a position to do something about this conviction. He became deeply involved in fundraising efforts, corresponding with various Anglican clergyman and members of the Society for the Promotion of Christian Knowledge (SPCK) in London, an Anglican voluntary society trying to start a Protestant version of the *Sacra Congregatio de Propaganda Fide* (founded 1622).[21] They were excited about the potential of Jablonski's efforts to help them overcome the growing status of dissenting religious groups in England.[22] Several of the first SPCK members, including the Royal Society's curator of experiments, Frederick Slare, became key promoters of the Orphanage.[23]

Leibniz became Jablonski's ally. Together, the two men urged Friedrich I to preside over a series of formal proceedings in which several experts would gather to discuss the possibility of moving beyond toleration in

Brandenburg-Prussia, including the kinds of institutions (a universal seminar, perhaps?) and techniques they would need to put into place in order to make this possible. In a text he coauthored with Gerhard Molanus, a theologian and mathematician who had studied under Calixt, Leibniz lamented that attempts at mutual toleration or "virtual union" were ineffective, making the condition of the "Evangelicals" (his term) worse by persecutions and secret oppression.[24] Shortly after this appeal, Benjamin Ursinus reported:

> His Royal Majesty is quite agreed that we should work for union and not for so-called tolerance—a union by which the unholy Schism might be terminated and by which one party and the other might be able and willing to use with clear conscience the worship and Communion of the whole. His Royal Majesty therefore thinks that it is necessary to abolish the party names, Lutheran and Reformed, and henceforth to call both churches by the single term Evangelical.[25]

At the request of the elector, Jablonski penned a response to Leibniz and Molanus's text, which Ezechial Spannheim personally delivered to Leibniz while on official business in Hannover. In it, Jablonski told Leibniz that Friedrich I supported the initiative, and precisely the kinds of interfaith discussions that Calixt had called for could commence.

At the same time as he was getting the Berlin Academy up and running and corresponding with Francke, Leibniz was also traveling to Helmstedt to talk about confessional union with two Calixtean theologians whom he had helped acquire positions at the university there: Fabricius and Johann Andreas Schmidt. Both of these men became early Berlin Academy members as well. While in Helmstedt, Leibniz convinced Calixt's son to edit and publish new editions of his father's irenical works. He had finally found a ruling prince who wanted nothing more than to formalize policies of confessional union. Leibniz returned to Hannover, where Schmidt and Fabricius joined him in the summer of 1698. Recalling that moment shortly before his death in 1716, Leibniz said that the two had come to see "whether there were any means of coming to better understandings between the two Protestant Churches, of which the theologians of Brandenburg-Prussia and Brunswick, who were always the most moderate in the Empire, might lay the first foundations."[26]

A small interfaith council convened that included Leibniz, Jablonski, Fabricius, and Schmidt. The group agreed that variances among Protestants, in ritual and creed, had to be done away with and the names of various Protestant groups needed to be completely abandoned. They also agreed that the key to resolving the problem involved not only overcoming the *appearance* of dissonance; it also involved helping others see that behind the differences embedded in each confession's ritual performances lay a single set of precepts articulated in the Bible. Jablonski returned to Berlin in the wake of the deliberations confident that progress had been made. Now, people only needed to be led "to agree that in the doctrines which divided them there was really no essential divergence," he wrote.[27] Teaching them how to see this through new pedagogical strategies was at the heart of the irenical turn.

These negotiations stagnated and ultimately faltered after a different irenical initiative began in 1699 that many, including Leibniz, disapproved of because of how it was undertaken.[28] To make matters worse, in 1702 Friedrich I grew impatient with the slow pace of deliberations and ordered that his evangelical bishop, Ursinus, call a *Collegium Irenicum*; some began to suspect that the crown's irenical aspirations might actually be better understood as "a foil, designed to achieve 'not so much peace . . . between the two Churches as the introduction of Calvinism into the Brandenburg territories.'"[29] By throwing his support behind these initiatives, Friedrich I stood to gain more acceptance of Reformed principles and groups he had invited to Berlin; he also hoped to legitimize a new form of "theocratic absolutism" in the process. In fact, he liked the idea of constructing an absolutist state whose civil governing structures were rooted in a non-confessional version of divine law.[30] Contingent upon a union of confessions, this form of absolutism was largely a reaction against Samuel von Pufendorf's system of universal rights, which Leibniz and many others believed was fundamentally flawed. Leibniz's metaphysical conception of justice (*justia est caritas sapientis*) went against Pufendorf's antimetaphysical or civil conception of natural law. Pufendorf's version of Grotian natural law drew from Bodinian sovereignty doctrine, "a native German 'statist' political science, a secular cultural and political history and German *Staatsrecht*—the political jurisprudence derived from the imperial legal system in particular."[31] His political theory helped legitimize the separation of sacred and civil authority, laying a foundation for a multi-confessional state—not the opposite. In response, Leibniz developed a "Christian-rationalist" version of natural law that appealed to many of his colleagues in Berlin and Halle. He wanted to shore up Christian natural law against what many understood to be Hobbes's, Grotius's, and Pufendorf's secularizing (of ethics and politics) projects, laying a foundation for a monoconfessional state—a universal religion and politics.[32]

In short, the irenical turn in Brandenburg-Prussia was about the implementation of a "Christian rationalist" form of universal jurisprudence. And when Friedrich I died in 1713 and his son Friedrich Wilhelm took over, he took this mandate even more seriously than his father did.[33] The emphasis then was on creating (with the help of the Orphanage) a state-sanctioned theophilosophical system linked to rituals foreshadowed in Old Testament accounts of the ancient Israelites, who would serve as models for ethics, politics, and expressions of community. The successful implementation of an irenicist agenda hinged on the institutionalization of techniques that would help create in people a heightened ability to appreciate the wider significance of these particular models.

Observing with the Inner Eye

Anschauung has long been linked to Kantian philosophy; however, the term was in use in the German states in the seventeenth century. One finds it, for example, in the works of Jakob Böhme, a Lutheran mystic who included a section entitled "How the Soul can be made to arrive at *Anschauung* and Obedience of God" in his nine-volume *Way to Christ*, published in 1682.[34] Difficult to translate precisely into English, *Anschauung* connotes an overtly visual process of experiencing and understanding—from the senses to the mind via intuition. One used this word to designate a combination of intuitive perception, contemplation, and conception. It also referred to the cultivation of attention and the various forms of experience one needed to embrace in order to properly understand something. *Anschauung*'s active form is *anschauen*, to observe: a sensory, cognitive, and emotional act imagined as the culmination of efforts to see with the outer and *inner* eyes.[35] This heightened ability to observe came only at the end of one's practice of piety after one had become adept in the three basic steps of meditation: seeing (*sehen*), reflecting/contemplating (*betrachten*), and observing (*anschauen*).

The Jesuit Hermann Hugo popularized these steps in his *Pia Desideria* of 1624, introducing a key link between pious desires or feelings and seeing and intuiting.[36] Comenius also described the eye of inward seeing (*das Auge des inneren Sehens*) as the spirit of understanding.[37] He believed it could be awakened by demonstrations and required light, the cultivation of attention (the strategic application of reason), and objects, which stimulated the senses. Comenius argued that the procedure for seeing with the outer eyes was fundamentally deductive, whereas the procedure for seeing with the inner eye was inductive.[38] Still, it was in the inner eye that many contemporaries believed perception and cognition converged, generating inexplicable forms of intuitive comprehension.

Amid the rush of interest in the state-supported irenical turn, Spener borrowed the phrase *Pia Desideria* from Hermann Hugo and used it to introduce a devotional book by Johann Arndt, a sixteenth-century Lutheran theologian, devotional writer, and chemist. In his very popular *Four Books of True Christianity*, Arndt drew upon a long tradition of hermetic and alchemical writings as he argued for the power of love as a physical force and described the heart as a mirror, an instrument that receives light directly from a divine source and directs it toward transforming something else.[39] Spener's *Pia Desideria*, or *Heartfelt Desires for an Improvement of the True Evangelical Church Pleasing to God*, was about the importance of verifying faith through actions that were visible, noticeable.[40] His central concern, following Lewis Bayly and many others, was with bringing "the head into the heart" and teaching his followers how to forge practices of piety—and transform the world—through a combination of meditation, erudition, and social activism.[41] In his own writings, Spener pursued this combination of love and illumination as the logical culmination of these practices.[42] He argued that the presence of love in a person's life would make itself manifest through visible signs.[43]

As mentioned previously, devotional literature as a genre proliferated in the German states throughout the seventeenth century as contemporaries responded to this idea that a return to apostolic Christianity might eliminate conflict and confessional difference. Devotional images, or emblems (*Sinnbilder*), became increasingly important parts of the practices of piety promoted in this literature. Protestant groups promoting renewal increasingly advocated contemplating devotional images—or embracing a set of practices already promoted in Catholic institutions.[44] The radical Pietist theologian, poet, and pastor Gottfried Arnold created his own collection of poetry called *God's Rays of Love* with this in mind.[45] Images of individuals with burning hearts or who were experiencing moments of divine illumination appear as emblems alongside the poems, which Arnold intended to be sung as hymns (figs. 7 and 8).

Scholars of German Pietism have tended not to emphasize the importance of observing—including contemplating emblems and devotional images—for the community of Pietist theologians associated with the Orphanage. At least one reason for this stems from long-standing concerns about the uses of devotional imagery among early Lutherans, who often pointed to the dangers of images. By the early sixteenth century, devotional images had become standard features of Catholic worship services, and many contemporaries believed they could "enter into an affective personal relationship" with those who observed them.[46] Early Protestants were critical of this and generally portrayed images as disruptive. Martin Luther admitted that he, like most people, could not help creating images

7   Frontispiece showing a burning heart, from Gottfried Arnold, *Göttliche Liebes-Funcken*,
     vol. 1 [1701, image from 3rd ed., 1724], engraving. Credit: Staatsbibliothek Berlin [Ct 2627],
     courtesy of Art Resource, NY

(*phantasias*) in his mind that helped him to grasp "theological truths";
however, he was concerned with their proliferation in the church and
with the widespread conviction that they stood in for divine messages
contained in the Bible. In response, he championed direct engagement
with the authors of the Old and New Testaments.

Although Francke advocated intensified engagement with these authors through careful reading, he also advised his students to spend a

8   *Amen im Herzen*, from Gottfried Arnold, *Göttliche Liebes-Funcken*, vol. 1 [1701, image from 3rd ed., 1724], engraving. Credit: Staatsbibliothek Berlin [Ct 2627], courtesy of Art Resource, NY

considerable amount of time and energy "in rapt contemplation" so that the holy spirit would *"open [their] eyes to see* the wonder in God's law."[47] He was interested in the contemplation of devotional images as a form of spiritual exercise and a means of generating "illumination" (*Erleuchtung*). He offered his thoughts about illumination, love, and contemplative seeing in his sermons and other writings, where he stressed the importance of cultivating the inner eye, or "the eye in people," as a way of generating wisdom.[48] In his *Philanthropia Dei* (1705) and *Philotheïa* (1706), he also provided instructions for how to recognize the love of God toward men and how to cultivate a burning or "seraphic love" of God by observing beautiful, virtuous, and inspiring models (for more on this specifically see chapter 4).[49] His stance is comparable to Robert Boyle's in his *Seraphic Love, or Some Motives and Incentives to the Love of God* (1659).[50] In this text, Boyle defined "seraphic love" to be "of a flaming Nature"; he explained that it was a kind of love

whose imployment, mentioned in the Evangelicall Prophet's Mysterious Vision, sufficiently poynts at the divine Object, to which the flames, that warm them, aspire and tend. . . . How much must a due contemplation enamour us of that divine, and, though refulgent, yet ever more and more discoverable object; where attention and wonder, still mutually excite and cherish each other?[51]

In 1708 the Orphanage published one of the first German translations of Boyle's *Seraphic Love*.[52] It appeared shortly after Francke began corresponding with Boyle's assistant and SPCK member Frederick Slare, who was involved in helping send boys from London to be educated in Halle.[53] Francke also continued to explore his interest in love, illumination, and the "inner eye" in later sermons, including one called the *Sagacity of Children of the Light* and another called *The Eye of Faith*.[54] *The Eye of Faith* is a clear nod to Raymond of Sabunde's *Oculus fidei*, a fifteenth-century treatise promoting the reconciliation of theology and philosophy that had been edited and reissued by Comenius in 1661.[55]

In Halle, efforts to better understand the power of "seraphic love" and the "inner eye" were linked to a pressing interest in trying to explain the relationship between passions, desires or affections—associated with the lower faculties or sensitive soul—and cognition, associated with the higher faculties or the rational soul.[56] By about 1700, the basic problem was this: if one accepted that the body was best understood as a kind of machine, then the very idea that there were embodied forces or tendencies that helped people perceive things or even make decisions was hard to believe. Mechanical philosophers were arguing that all matter, including the body, was inert and lifeless, which meant that an "inner eye" would have to reside in the mind, if it existed at all. If one believed it was possible to truly apprehend with this inner eye, then he or she might be wary of the idea that passions, desires, and affections were not located in the body. Christian Thomasius was very wary and attempted to explain why in a treatise on natural philosophy he wrote in 1699, wherein he rejected a mechanistic view of nature and explained his concern that Cartesians placed too much confidence in the powers of "mind over matter."[57] He also argued that this was the problem with orthodox Lutherans: they overly stressed the powers of the mind by continuing to insist upon the importance of logic for the study of philosophy at the expense of attending to people's real emotional needs.[58]

Thomasius believed that a person's desires or affections had a determining effect on cognition. Because of this, he felt it was important to find a way of ensuring these desires were pious (*pia desideria*) and oriented toward the good. In his opinion, the only way to do this was first to "attain the state of Christian faith" through an authentic conversion experience.[59] Thomasius wanted more of what he called "rational love" in the world, but he believed it had to start with an "honest passion" located in the body that then impacted the mind—sometimes producing forms of illumination or enlightenment.[60] Leibniz disagreed. He spoke out in opposition to Thomasius's ideas, explaining:

I am not in agreement with him when he makes love come *before* enlightenment [*la lumière*]. . . . Enlightenment is our passion, love is the pleasure which results from it and which consists of an action upon ourselves, from which comes an effort to act thereafter on others, to contribute to the good insofar as it depends on us.[61]

Leibniz insisted it was possible to improve the mind through work on the self and through exercise, which would lead to an improvement of the quality of one's affections and desires. An avowed proponent of a "Christian rationalist," mechanical worldview, Leibniz was also deeply invested in efforts to revise popular understandings of emotions as embodied (i.e., as "forces of nature") and to relocate them in the mind.

Given Francke's own interest in *philanthropia dei* and the inner eye, he may very well have been inclined to agree with Thomasius. But in the end, he sided with Leibniz against Thomasius. It did not help that Thomasius carried the implications of his argument as far as he did. Shortly after the founding of the Orphanage, Thomasius asserted that its schools and the Pädagogium inevitably would not have the kind of effect that Francke and so many others believed they would have.[62] He insisted it was not possible to generate piety or to turn young people into morally upright adults through the kinds of educational programming the Orphanage was implementing without ensuring that all those who entered it had already undergone a genuine conversion experience. In response, Francke and other members of the theology faculty accused Thomasius of "examining questions he was not competent to discuss."[63] After all, he was basically undermining the aims of all those involved with the Orphanage.

Again, Francke probably appreciated Thomasius's concerns about generating true, rather than merely superficial, forms of piety; however, we cannot forget that he was on the defensive against those who wished to associate the Orphanage with religious enthusiasm in the wake of the Leipzig controversies and with Francke's associations with ecstatic women in Thuringia.[64] He had a great deal of latitude as director of the Orphanage; yet at the same time he needed the support of the king and his allies at court in Berlin. They would be less inclined to support him if he could not somehow repair his reputation as an outspoken and disorderly *Schwärmer*. "One really ought to think about what kind of people Francke and his accomplices in the Pietist enthusiasm are," Johann Heinrich Feustking (writing under the pseudonym Hieronymous Bahr) noted as late as 1709.[65] Feustking believed Francke and his followers were really dangerous and ecstatic Calvinists who had been given license to institutionalize false and fanatical teachings.[66] Francke was aware of these kinds of criticisms and their potential to damage the new community's reputation; he had to defend the integrity of the Orphanage as a "universal seminar" and a cutting-edge place for research. Those involved in the irenical turn were relying on him to standardize strategies that would help young people learn how to look past forms of confessional difference and intuitively apprehend universal truths. This was going to take time, special tools, exercise, and discipline.

Inclusionary or Exclusionary Eclecticism?

The irenical turn generated an interest in strategies or tools that might help create a thoroughly historicized and universal system of pious natural philosophy. The *via eclectica* held appeal as a tool because of its emphasis on evaluating and assembling evidence. Eclecticism was a formalized way of studying and synthesizing observations (acquired through correspondence, conversation with reliable witnesses, or direct experience), descriptions, theories, opinions, and explanations in order to better understand a particular phenomenon. In Halle, Altdorf professor Johann Christoph Sturm was considered eclecticism's first major promoter and had been fond of arguing that it was the mark of an educated and freethinking individual.[67] It involved a choice to see the good in as many perspectives as one could bring to bear on a particular issue as possible—and it freed one up from having to choose a particular philosopher or "school" to belong to, thereby contributing (many hoped) to the elimination of sectarianism.

Sturm's former student Johann Daniel Herrnschmidt promoted eclecticism in Halle's Orphanage.[68] We can get a sense of how he understood it in a short essay he wrote called "The Proper Boundaries of Natural Philosophy." Here he described

a certain way of researching, named either the *via eclectica* or *electiva* which helps one overcome the limitations of human understanding. Above all it has to be palpable, so that in something as expansive as the study of nature a single man does not have to be right about everything and always know the reasons why he can neither blindly rely on Democritus, Plato, Aristotle, and some old wise saying or the phrases of the new philosophers but rather, with time, will be able to test the increasingly clearer qualities of natural things and from these tests assess the reasonable conclusions that emerge from them.[69]

The starting point for this special way of researching was a desire to remain open to a variety of perspectives and awareness that it was impossible for any one person—one philosopher, one observer—to study everything. Because of this, Herrnschmidt argued, "one should protect freedom of choice."[70]

Eclectics cultivated openness toward a variety of opinions and perspectives because it was more likely that a single individual was right about some things but not others. He explained:

In recent times there have been many new discoveries and especially chemical experiments, and the microscope, telescope, air pump, barometer, thermometer etc. have made possible a more exact study of the inner form of bodies, and distant bodies in the universe . . . and many new philosophers have become quite successful; still it does not follow that as a result of this one should believe everything one man says or expect the truth from one person.[71]

It was important to take a little bit from one source and more from another, Herrnschmidt argued, and to decide for oneself, based on the available evidence, what was believable and what was not. Even better was to go straight to things in the world, to gather firsthand observations, conduct tests, and come to one's own conclusions. This way, one would gradually come to understand the larger philosophical issues at stake without feeling compelled to decide which philosopher or school of philosophy one agreed with first and then to see everything through this lens.[72]

In this same passage, Herrnschmidt argued that no other method was more natural and more capable of helping correct the imperfections of human understanding than the one that begins by carefully noting the "effects and accidents of creation that can be felt with the help of senses."[73] He insisted upon the importance of gathering *Observationes* by seeing, hearing, tasting, smelling, feeling, and, most importantly, by improving or extending the power of the senses through the arts,

so that one uses the microscope, which presents the naked eye with invisible small things, or the telescope, which allows one to see things clearly and distinctly at a distance that would be otherwise unrecognizable. This kind of observation of natural things through the senses is standard operating procedure in physics and is the foundation upon which all studies of nature must be built.[74]

The kind of observation that worked hand in hand with experimentation, which he said had become "standard operating procedure" in physics, played a crucial role in Herrnschmidt's own understanding of the *via eclectica*. He recommended consulting Jena professor Hermann Friedrich Teichmeier's *Elements of Experimental Natural Philosophy* and the writings of J. C. Sturm, Robert Boyle, and others for examples of how performing experiments could provide one's reason with keys for improving one's understanding of natural things.[75]

In many ways, Herrnschmidt's description of eclecticism is very Baconian; indeed, the popularity of eclecticism helped create a rationale for the introduction of empirical methods and technologies in this part of Germany. However, being an eclectic also meant being inclusive and open to opposing points of view, an aspiration that was consistent with Tschirnhaus's and Leibniz's efforts to reconcile the growing rift between empiricists and those who adopted rationalist approaches to knowing the world. One who pursued the *via eclectica* was not supposed to rule out certain perspectives entirely; the idea was that everyone who had been deemed gifted or able enough to participate (by their teachers and mentors) in ongoing discussions had at least something to offer. Their perspectives deserved some thoughtful consideration and, at the very least, respect—even if one decided that most of what they had to say was wrong or misguided.

In the introduction to his *Rational Thoughts about God, the World, the Human Soul, and Everything,* Christian Wolff noted that despite saying they were looking for the best in all perspectives, people often did not practice eclecticism in this way.[76] He worried his colleagues at the university were promoting a form of eclecticism that involved excluding or rejecting certain voices entirely. He called them on this and claimed that he, on the other hand, offered "the right method" for eclectic philosophy:

One can see again that I do not advocate anything sectarian . . . but rather that I am searching for ways to hold onto all that is good, and I will find it wherever it is. . . . This, I think, is the right method for an Eclectic Philosophy, or a way of knowing the world that does not belong to any flag but rather tests everything and retains the things that can be joined together with reason, or that let themselves be brought together in a system of harmony.[77]

In this passage Wolff stressed the importance of making reason the main tool of assimilation: it was what made it possible for the practitioner of eclecticism to appreciate what was good or valuable in a particular point of view. This kind of assertion was very closely connected to the ongoing debates surrounding where exactly emotions, or, more specifically, feelings of affection or appreciation, originated—in the body or the mind.

By publishing his *Rational Thoughts*, Wolff was not only making public his concerns about an exclusionary form of eclecticism that was becoming the norm in Halle; he also endeavored to create a rational framework for understanding God, the world, the soul—well, everything. He seemed to be valorizing philosophy at the expense of theology, even challenging the authority of the theology faculty itself. Several professors in Halle made their objections public, including a fellow philosopher at the university named J. F. Buddeus.[78] Like many of his colleagues, Buddeus accepted the lingering institutional hierarchy and aspired to move into a theology faculty, still the most important faculty in a German university. He eventually achieved this goal, accepting a theology professorship in Jena in 1705.

Buddeus had all kinds of problems with Wolff and his *Rational Thoughts*; he thought Wolff had too much faith in a mechanistic worldview and the powers of human reason. He also worried that if a practitioner of eclecticism made human reason the primary engine of choice as he assimilated information and points of view, this would mean that no moral principles would necessarily inform these choices.[79] These principles would become a secondary priority—or be ignored entirely. "Ultimately a man is worthy of the name and title of eclectic," Buddeus wrote in his *Eclectic Philosophy* (*Philosophia Eclectica*),

who accurately devises principles for himself by the examination of things themselves. Let him *according to the standard of his principles* select all those arguments which he reads in the work of others, *appropriating for himself those which suit these principles; yet rejecting those that cannot be reconciled with them.*[80]

This powerful passage conveys the alternative understanding of eclecticism Buddeus promoted. Contrary to using reason to select arguments, he encouraged the practitioner of eclecticism to make choices "according to the standard of his principles." If one's moral standards did not permit themselves to be reconciled with a particular argument or perspective, then it was permissible, according to Buddeus, to reject it entirely.

What if an aspiring eclectic was not entirely aware of the moral standards or principles he or she should be using to make decisions about what perspectives to reject and what to retain? Buddeus argued that the key to grasping this standard was studying the "history of morality," which he was in the process of writing.[81] Included in this was the study of the history of philosophy, which he noted was so often neglected. He believed more focus on this would draw attention to the "defects of previous schools" and to aspects of new philosophies that were simply repackaged versions of older ideas.[82] In his *Introduction to the History of Hebrew Philosophy*, he pointed to Comenius as a philosopher who understood the importance of historicizing philosophy too.[83] He tried to link various components of Comenius's pious natural philosophy, which was already harnessed to irenicist aspirations and politics in Brandenburg-Prussia, to eclecticism— but more specifically the form of exclusionary eclecticism he preferred.

The irenical turn in Brandenburg-Prussia was supposed to generate a theocratic system of laws, practices, and technologies foreshadowed in the rites of the ancient Israelites. To understand these rites meant studying history—a subject often pursued by theologians. By 1700, the study of history had faced a series of attacks by Descartes, Spinoza, and many others, who argued that human testimony—past and present—was unreliable and that accounts of past events, even those contained in the Bible, could not get one closer to apprehending truth. These attacks had forced those interested in writing history to reevaluate their practices, but they did not stop people from doing history. Quite the contrary: they sparked an interest in writing "cosmogenies," or conjectural histories of the earth (see chapter 5), in devising strategies for more critically evaluating evidence from historical documents and in treating historical artifacts and monuments "as especially unbiased sources" for understanding particular places and moments in the past.[84]

Comenius's natural philosophical system privileged the study of physics and biblical history over mathematics and logic; it started with a discussion of God and then considered the world's "pre-history" and the moment of creation via a literalist reading of the Bible. Comenius insisted that the world was created in six days and would come to a definite end.[85] This was in stark contrast to what Wolff later argued (following Leibniz), namely that the world was eternal.[86] After reflecting on this history, Comenius introduced three principles underlying everything: matter, spirit, and light.[87] He explored the causes of motion and metamorphosis, returning to Aristotelian ideas about motion as an "accidental" quality of bodies but ultimately concluding that both motion and metamorphosis were better understood as tools (*Werkzeuge*) of nature.[88] Finally, he described the "qualities of the things of the world," which were perceivable through the senses. He offered a seven-part guide to understanding the layers of these things and their qualities—namely ether, heavenly bodies, things in the atmosphere (fog, clouds, rain, snow, etc.), minerals, plants, animals, and humans.[89]

As Orphanage administrators considered if and how they were going to teach Comenius's natural philosophical system in their schools, they had to face up to several things. First, it was outdated, originating in the 1630s. Second, they would have to decide how literally to interpret biblical accounts of a six-day creation and other events like the Great Flood. And third, they would have to prepare their students to respond to the criticism of those who very likely would continue to insist that the study of biblical history combined with physics was a less reliable way of helping young people systematically apprehend the truth—especially truths about the natural world and its operations. Mathematics did not play a central role in Comenius's system, yet Francke wanted to ensure that young people in the Orphanage were trained in mathematics, including geometry. This was one of the reasons he was so interested in collaborating with Tschirnhaus and Leibniz in the first place.[90]

When the Orphanage was first founded, administrators had engaged selectively with Comenius's writings and the tenets of his natural philosophy—most directly with his more general admonition (in the *Janua linguarum reserata* and *Orbis Pictus*, for example) that "true philosophy should be grounded on a real empirical language."[91] In 1702, Francke noted that in the Orphanage schools student teachers used a short text first published in Gotha "wherein the principles of the most important and useful sciences are shortly summarized."[92] This text was probably Andreas Reyher's *Short Lessons about Natural Things*, which Reyher had derived from Comenius's pedagogical writings.[93] Reyher was the educational advisor of Duke Ernst the Pious in Gotha, which was one of the first German cities where educators actually applied Comenius's recommendations to their reform initiatives. Francke had attended school in Gotha, where his father was a political advisor to the duke, and was quite familiar with these efforts. In his correspondence with Comenius's grandson, Jablonski, Francke discussed ways they might get more of his grandfather's manuscripts translated and published.[94] He acquired some of these manuscripts, and the Orphanage printed summaries of Comenius's *General Discourse on the Emendation of Human Affairs* and his *History of the Bohemian Brethren* in 1702.[95]

However, Francke must have known there were limits to Comenius's natural philosophical program and that it was not possible to promote eclecticism and simultaneously allow only one perspective or philosophy to dominate the curriculum of the Orphanage schools. His descriptions of these schools always emphasized the variety of prescriptions he was bringing to bear on them, and he consistently stressed the importance of teaching young people mathematics. Even in the vernacular schools, Francke wrote in 1702, young people learned physics, arithmetic, and geometry "so that when they are apprenticed in a particular craft they will have at least some knowledge of the useful sciences, which are very important to the common way."[96] In his early pedagogical writings he praised the ancient Greeks,

who offer a beautiful example in that they led their children to *Mathesis* at an early age, which is advised by certain mathematicians today as well. Of course, one should not forget that *Mathesis* is too difficult for children. But if one leaves the most difficult things pertaining to demonstrations and calculations for riper years, one will still find enough in mathematics to inspire children with passion and pleasure—without overwhelming them.[97]

Especially young children ought to be taught astronomy, Francke noted, because it provided the perfect opportunity to teach them to recognize the size, quantity, and order of heavenly bodies.[98]

In 1702, a typical *Pädagogium* pupil spent many hours a week studying Latin (a bit of Hebrew and Greek too), geography, history, and theology, in addition to arithmetic and *mathesis*, with the help of Tschirnhaus's *Basic Guide*. In history classes pupils learned about a variety of important people from the past, from several cultures and epochs; student teachers taught history by showing young people pictures of these people, "one after the other . . . and letting them remark on what period or century each one lived in and what they did."[99] This was supposed to make it easier for them to apprehend the synchronism, or relationships connecting seemingly distant figures and events, and to remember them and their respective epochs.[100] In their periods of recreation, pupils sang songs together and learned the catechism; practiced astronomy; talked with their teachers and one another about the "beauty of manners"; engaged in exercises involving the organization's collection of curious things, including observing experiments; attended anatomical demonstrations; perused books and catalogs in the Orphanage bookstore; and went on botanical fieldtrips. They also visited craftsmen and turned wood.[101] But there was no single system of physics or natural philosophy that structured or justified the ordering of these activities.

A revised description of the *Pädagogium*'s curriculum that appeared in print in Halle in 1721 shows just how much things had changed. The author, *Pädagogium* director Hieronymous Freyer, emphasized that Buddeus's *Elements of Instrumental Philosophy* was then required reading for all student teachers. By this time Buddeus was teaching in Jena's theology faculty, and his writings had become increasingly popular; at least twenty-five editions of his two-volume *Elements of Instrumental and Theoretical Philosophy*—referred to simply as his *Eclectic Philosophy*—were published in Halle between 1703 and 1727.[102] In these books, Buddeus tried to revive a literalist biblical historical narrative about a pious natural philosophical system that would effectively extend the system described by Comenius in his *Natural Philosophy Reformed by Divine Light*.[103]

Freyer noted that especially the first two chapters of the *Elements of Theoretical Philosophy*, in which Buddeus described the qualities of animals, plants, and fossils, now served as the starting point for exercises pupils undertook during their recreational periods. A carefully historicized system of mosaic physics now structured these activities. Whereas in 1702 practicing astronomy—a form of applied mathematics—was listed as the second-most important exercise that pupils undertook (second only to singing songs and catechization), by 1721, it had moved into the last category. The first category or set of activities was now called "Preparation for Physics and the Bible."

Reinvigorated and promoted by Buddeus, Comenius's system of mosaic physics assumed a new importance. Now student teachers privileged field trips to visit artists and artisans and short lessons about the properties of animals, plants and trees, metals, stones and minerals, the earth, water, air, fire and meteors, *Oeconomie* and medicines for study during recreational periods.[104] These lessons culminated with an explanation of the Temple of Jerusalem (see chapter 4).[105] Then pupils were given time to practice the "mechanical disciplines" by turning wood, making instruments and figures out of cardboard, and polishing glass.[106] They further explored the "disciplines of physics" by botanizing and observing dissections and experiments.[107] Lastly, they undertook activities associated with the "disciplines of mathematics," specifically astronomy, music, drawing and calligraphy.[108]

Evidence that teachers were using Tschirnhaus's *Basic Guide* disappeared, although Freyer acknowledged they relied on excerpts from Wolff's *Beginner's Guide to the Mathematical Sciences*.[109] Teachers still taught mathematics in the *Pädagogium*, especially geometry, trigonometry, and algebra. However, Freyer stressed that they did so more carefully or deliberately—that is, in a way that would help young people better understand it as a real and useful science. Freyer stressed that pupils were more frequently taken out to the botanical garden, for example, where they measured the length, width, thickness, and height of different bodies.[110] He described how instructors worked more vigilantly than they had before to ensure that their pupils understood that points, lines, and curves, for example, were not abstract concepts or pure "rational essences" (*rationales*), as Tschirnhaus had described them; rather, they originated *in* the world. When asked how they came to understand why lines exist and what they are, Freyer said his pupils knew this because

(1) they had looked at [*ansehen*] and observed [*betrachten*] the line;
(2) from these observations they had apprehended the peculiarities that would allow them to distinguish it from other things; and
(3) they then considered for themselves how or why it had come into existence.[111]

This, Freyer noted, was an ideal way to help them develop clear concepts of things and to begin to recognize truths about the world.[112] Accustomed to this set of investigative procedures, they would then seek out other definitions, histories, and perspectives on the things they had observed.

By 1721, *Pädagogium* pupils were spending more time studying theology, biblical history, and geography than they did before. Regarding the teaching of geography, Freyer wrote that "all four parts of the world must be covered: Germany and Palestine above all things, however, so that the young people can make uninhibited progress in understanding their fatherland and biblical history."[113] Freyer noted that no one was allowed to learn history who did not have some training in geography and that now only a "universal history" of the Old and New Testaments was taught. Whereas in 1702 pupils learned about a variety of historical figures, now the main focus was Old Testament fathers: Adam, Noah, Solomon, etc. Highly talented pupils could receive some introductory lessons in medicine, law, and even philosophy (at least six hours a week) to help prepare them for related collegia at the university. Yet again Freyer noted that teachers "used the writings of Buddeus," not Wolff, in their philosophy lessons.[114] These lessons, he continued, were offered in the interest of caution "so that the character of the youth was not corrupted by the study of philosophy, which unfortunately occurs all too often."[115]

These curricular adjustments unfolded in response to mounting tensions largely linked to concerns about Wolff and his brand of eclecticism, which stressed the importance of allowing reason—not moral principles—to guide one's efforts to synthesize information and competing perspectives. More generally these tensions stemmed from Wolff's challenge to the hierarchy of disciplines in German universities and a growing sense of unease among Halle's theologians. As Thomas Howard writes, "Wolff insistently rejected the dependence of philosophy upon theology, arguing instead that the intellectual currency of the philosophical faculty, human reason, must be that of the higher faculties."[116] He parted ways with those who believed theologians ought to continue to direct the study of natural philosophy.[117] One powerful implication of Wolff's increasingly aggressive assertions was that the Orphanage should not necessarily be the exclusive domain of the theology faculty.

Although there is very good reason not to trust Wolff's account of his own expulsion, it is interesting to note what he reportedly said the "cause of the whole thing was" later in his life.[118] He claimed Francke had become convinced that he could not turn someone into a Christian by teaching them geometry and increasingly disassociated himself from the idea that mathematics was an effective form of spiritual exercise. As Wolff explained it:

Among the theologians, Francke held the opinion that he could not turn someone into a Christian who studied Euclid. He expressed this in opposition to the honorable Professor of Mathematics Rudolph from Erfurt, who advised him to put Euclid in the hands of all the youth of the Orphanage and the *Pädagogium*; however, he could not really give a reason for it, only said he knew it from experience.[119]

If this is really what Francke said on the eve of Wolff's expulsion from Halle, he had made statements to the contrary before this. In one of his earliest letters, Tschirnhaus had written to Francke that much of what he loved about the Orphanage was that the children who trained there learned geometry. He appreciated the fact that Francke was willing to apply his suggestions, which involved finding a middle way between empiricist and rationalist approaches to knowing the world. But as time went on, Francke and other administrators had numerous opportunities to reflect on just how effective their techniques really were and made adjustments where they felt they were needed. They also had ample time to reflect on the kind of eclecticism they preferred. Whereas Francke once had brought a variety of perspectives to bear on the Orphanage's educational programming, by 1721, one perspective—Buddeus's exclusionary approach to the *via eclectica*—came to dominate all the rest.

FOUR

Models and Conciliatory Seeing

In perspective lies the model
Before the eyes clear, distinct and bright
Just like the one in Halle's Orphanage
Where all the parts are, above all,
Erected according to the book.
JOHANN JACOB SCHEUCHZER, *KUPFER-BIBEL*

Three-dimensional models have long been a part of attempts to resolve tensions between the physical world and one's ability to attach meanings to it. The status of the assimilatory work models do has always been in flux, and the legacy of this instability is still with us today. It is perhaps best evidenced by a model's capacity to connote several things at once: a miniature building, a beautiful human being, or even a way of thinking (i.e., a mental model). In the early modern period, material models of machines and buildings could be found in many cities. But they were not readily accessible to most people. Viewing them required access to a curiosity cabinet or an artist's workshop. By 1700, this was starting to change. Models were more openly displayed. They were also increasingly valued as pedagogical tools that fostered better ways of understanding new technologies and materials.[1]

Some early modern thinkers described models as explanatory devices. For example, fifteenth-century architectural treatises often pointed to the model's ability to help convince wealthy patrons to move forward with a building project. These things were useful to architects and engineers because they could help remedy potential problems while still in a planning stage. In the eighteenth century, models were increasingly described as demonstration devices that could help diffuse philosophical principles or propositions. In 1718, Leiden professor Willem 's Gravesande wrote a letter to Isaac Newton in which he said he was "obliged to have several machines constructed *to convey the force of propositions* whose demonstrations they have not understood."[2] In this case, model machines helped convey a proposition or bring abstract principles down to earth.[3]

J. C. Sturm's son, a mathematician and architect named Leonhard Christoph Sturm, defined models as "material ideas" that facilitated new ways of seeing, conceptualizing, inventing, and knowing.[4] When built to scale using the geometrical projection techniques of artists and engineers, models presented observers with a special combination of the three essences E. W. von Tschirnhaus had identified in the *Medicina Mentis*: the real, the imaginable, and the rational. They inspired a form of "conciliatory seeing" that was rational, affective, and rooted in the material world. Inside the Orphanage young people had the opportunity to practice this kind of conciliatory seeing by undertaking a variety of visual exercises. Such an aim required special techniques and tools. This chapter looks carefully at one of these tools, an eleven-by-eleven-foot model of Solomon's Temple that was placed on display in the Orphanage in 1718.

Since at least the third century, Solomon's Temple has been imagined as a divinely inspired, archetypical building, but it was not always clear what this meant. The historical Temple of Jerusalem, built by King Solomon, is described in the Old Testament books of Kings and Chronicles. The Temple was destroyed by the Babylonians, and the book of Ezra details its rebuilding under Cyrus and Darius the Great. Its complete reconstruction at the hands of Herod is detailed in Josephus Flavius's account of the Jewish-Roman wars, during which the Temple was destroyed completely for a second time. In the book of Ezekiel, the Old Testament prophet relates his vision of a future temple to members of his community then in captivity.[5]

Some architectural treatises traced a path of steady improvement from the beginning of time (Adam as the first architect) to ancient Rome.[6] And there were certainly prominent intellectuals in Halle who preferred to understand the significance of Solomon's Temple in this way. For example, Johann Samuel Stryk, a law professor at Halle University, argued in a 1702 disputation (*de jure Sabbathi*) that the impetus for the Temple's construction came originally from pagan worship traditions.[7] He pointed to the gaudy and unnecessary decorations, the gilded walls, and other useless items that he believed inhibited worship. Other treatises, Nicholas Goldmann's in particular, posited the Temple as a site of origins, arguing that the palace architecture of the ancient Greeks, Persians, Babylonians, and Romans was entirely derivative.[8]

Investigating and reconstructing the Temple's "sacred geometry" were popular scholarly enterprises in the early modern period. A wide range of scholars, often theologians, were interested in how ancient architects used geometry to engineer sacred space, making use of their special understanding of certain ratios or mathematical relationships, such as the Golden Mean.[9] Some also wanted to know more about if and how the historical Temple might have served the ancient Hebrews as a kind of administrative instrument.[10] Understanding how it helped build better states was an appealing prospect for territorial leaders, who supported early antiquarian investigations of the Temple in their realms. Its widespread currency as an object of inquiry made it easy to employ as a motif for new knowledge-making projects, such as the one that Francis Bacon described in *New Atlantis* (1627), which helped inspire the founding of new scientific academies.[11]

Those who witnessed the Halle Temple model participated in ongoing attempts to foster new modes of "collective perception" that were characteristic of this early eighteenth-century moment, including the founding of academies and "the cultivation and calibration of shared habits" for managing information.[12] As a site of initiation and assimilation, the model was an ideal tool for training eclectic observers.[13] Because it was intricate, impressive, and organized according to mathematical principles (*rationales* in Tschirnhaus's paradigm), it stirred the imagination (*imaginables*) and was supposed to inspire young people to want to know more about the original Temple. Because it combined and made visible a variety of expertise, one could use it to help individuals appreciate various efforts to reconstruct it; one might also reflect on the reasons there were so many different ways of portraying the Temple. Additionally, it brought observers as close as possible to a real historical artifact (*reales*) that might help anchor efforts to historicize a system of pious natural philosophy in Halle. In the first section of this chapter, I explore how Semler used the space of the Halle Temple model to make a combination of expert knowledge about the space visible. The second section uses several of Francke's pedagogical writings to consider how the Temple model served as a medium for awakening the "inner eye" and compelling observers to emulate the virtuous practices, individuals, and instruments they witnessed inside the miniature building. The final section explores the role of models and lessons in perspective offered in the Orphanage schools.

Reconciling Expertise and Awakening the Observer

When the architect and builder of Halle's exceptionally large wooden model of Solomon's Temple, Christoph Semler, finally unveiled it in 1718, he was already well known there as a model and instrument maker, teacher, and pastor. Semler spent most of his life in Halle, leaving only briefly to study at the University of Jena with Erhard Weigel. Semler strategically aligned his plans with those of the Berlin Academy of Sciences and others interested in making learning like play. He founded a new "school of the real" (*Realschule*) in his home in 1707, where he taught "useful sciences" to children using naturalia and instruments. "As much as possible," he wrote just before opening the school, "everything will be demonstrated in nature and in the presence of the objects."[14] His school was evaluated and approved by the Berlin Academy membership, but it only stayed open for about a year.

As a child, the philosopher Georg Friedrich Meier was one of Semler's pupils and boarders. In his autobiography he remembered that Semler was deeply involved in teaching mathematics in Halle and regularly collaborated with student teachers from the Orphanage.[15] Meier also described Semler as "a lover of mathematics" who built all kinds of models. He said Semler built

two models of the heavens, including the earth and the routes of the planets, and models of the city of Jerusalem, the Tabernacle of Moses and the Temple of Solomon that are all in the Orphanage's cabinet of curiosity; every day, three instructors were received from the Orphanage, who would each teach us for two hours a piece. In his house was also a mechanical fabrique. He worked on building earthly and celestial globes and we children helped him with this. He instilled in us, playfully, a love of mathematics this way.[16]

I will treat the relationship of these models to one another, including the heavenly spheres, in the last section of this chapter. For now I want only to point out that Semler's models were displayed inside the Orphanage and that the institution was also sending student teachers to Semler. Meier also made an indirect reference to Semler's ability to make mathematics *ludus,* by this point a trope linking Semler's methods to Comenius (the *Schola Ludus*) and to the efforts of Leibniz, Tschirnhaus, and Francke described in chapter 2.

Meier's friend and colleague, Samuel Lange, described Semler's interest in building a new school as an expression of his pious desires (*pia desideria*). In Lange's estimation,

Semler was a great mathematician. He was also one of the greatest mechanics of his time and, at the same time, a great astronomer. His knowledge of physics was exceptional. He was the first founder of a school that would become famous as a "school of the real" and published a small treatise about his suggestion for a mechanical school for all artists and artisans. At the time this was *pia desideria*. After it became realized, his name was not remembered. He said that models of all types of military and civil architecture, and the art of building ships, should be constructed. He said that models of wind and watermills should be made. The suggestions that were realized in the finished models that are now in the Halle Orphanage . . . all of these came from him.[17]

As Lange's account indicates, Semler was an influential figure who participated in the educational reform efforts that swept through Halle beginning in the 1690s. Francke's diary indicates that he and Semler conferred regularly about several matters, including plans to build a small workshop for producing globes on the grounds of the Orphanage.[18] In 1718 Francke reported to his patron Carl Hildebrand von Canstein that Semler had given him all the models from his *Realschule*. Canstein responded that he was overjoyed to learn that Francke had "received the mechanical school" and hoped that the entire plan of Herr Magister Semler would be realized.[19]

While founding and describing his new school, Semler also helped to popularize a scheme in Halle for assessing the quality of the knowledge that could be achieved with certain kinds of objects. The scheme drew from the Roman rhetorician Quintilian's characterization of evidence, which Stephen Gaukroger has demonstrated lay behind much of Descartes's call for the use of geometrical exercises to make manifest the relationships between ideas and the sensitive soul.[20] Models, Semler insisted, produced better, clearer knowledge than images and descriptions offered in books. As he explained:

There are different gradations of understanding to be acquired of visible things. For example, I can acquire understanding of the fruit of an Indian tree from

1. a description with words in a book. But this understanding is like death; it is like when one sees something at night. Now if the fruit

2. is described through word of mouth by someone who has seen it himself, then my understanding will become much more vivid because I can ask this person

more questions about that which I do not quite understand enough; this cannot happen with a book. Now if fruit of the foreign tree is presented

3. in a copperplate engraving, or
4. in a painting with vivid colors, or
5. in a model made of wax or wood, then the understanding is still much brighter and more distinct, or
6. in its natural state and laid out before the eyes. Then this, of all the aforementioned ones, is without question the highest degree of understanding and it presents the character with the best ideas.

Now when to this kind of ocular demonstration is added a vivid verbal lesson, then the understanding of the thing will sink itself more deeply into one's character. And this highest grade of the understanding of objects is intended by this project, which endeavors to show everything either in its natural state or in a model.[21]

In this passage, Semler communicated his conviction that all models possessed a peculiar ability to generate forms of understanding that were almost as powerful as what one acquired when one encountered the thing itself in nature.

He used a similar language of clarity and distinctiveness to describe the potential of his Solomon's Temple model. In the handbook he wrote to prepare observers for the viewing experience and to help them reflect on it later, Semler explained that he had built the model specifically for use in the Orphanage. He also endeavored to explain why his model was an especially valuable tool:

The descriptions and copperplate engravings one finds in the authors (of the holy scriptures), if they are truthful, are helpful for acquiring a clear idea of holy places and regions; but none of these produce the same kind of distinct or vivid idea as when one sees the entire Temple standing before the eyes in a material presentation and model. Here not only many sayings about the Temple that one only has a vague idea about become bright and distinct but the historical events, which took place in the Temple, are carried inside the character of a person, where they make a deep impression.[22]

Since the famous Temple of the ancient Hebrew King Solomon could not be visited "in nature," a wooden model of the Temple promised to offer observers the clearest, most distinct and vivid knowledge about the space. Semler argued that the model acted upon the observer; it generated sensible impressions whose quality would improve incrementally the better the model had been built to appear.

In its day, Semler's temple model was one of three large wooden temple models, each at least three meters in diameter and custom built to be observed systematically. Of course, images of Solomon's Temple abounded in the sixteenth and seventeenth centuries. The elaborate temple drawings of Juan Bautista Villalpando, Benito Arias Montano, and Isaac Newton are among the most famous and, even today, are often referred to as models or model drawings. While there is some evidence that Villalpando built a wax model to accompany temple drawings he presented to Philip II of Spain, he never described wooden models as things that could help people better understand the space than images or drawings.[23] In the 1640s, when a rabbi from the Netherlands named Jacob Jehuda Leon began exhibiting an intricate wooden temple model he had built himself, this changed.

Leon believed his wooden model would help people learn more about the Temple than they would by observing images of it. He profited from turning his model into a spectacle, charging visitors who came to see it a small fee for the privilege. By 1670, the size and intricacy of Leon's striking eleven-square-foot temple model had caught the attention of various members of the British Royal Society, who invited Leon, often referred to simply as "Templo," to come to England with it. Upon his arrival, Christiaan Huygens introduced him to the society's architect, Christopher Wren. Robert Hooke reportedly had extended conversations with Wren about Leon's model, which was received in London as a philosophical apparatus or instrument.[24]

There was at least one other large wooden temple model that had become famous in the years before Semler built his own. A man from Hamburg named Gerhard Schott built it in conjunction with the debut of opera there called the *Destruction of Jerusalem* (*Die Zerstörung Jerusalems*). Schott's model was approximately three and a half meters long but only about a half meter tall. Although it is not clear whether the model was actually placed on the stage during performances, it soon became quite popular. The eyewitness account of the brothers Uffenbach, who visited the Hamburg opera house during their *Bildungsreise* in 1709, offers a taste of how they experienced it:

The model of the Temple of Jerusalem, which stands behind the theater, is a piece of art that is certainly worth seeing. . . . The base is made out of oak; the outer walls and decorations are made of pear tree; the decorations, festoons etc. are made out of pear tree bark, which certainly look really nice. I also saw impressions of medallions. In this Temple there are also a total of 6726 pillars, which have the same amount of delicate capitals and were all constructed out of lead and crafted by a goldsmith. . . .

The whole work is entirely hollow and has many artfully made vaults. One can take all of it apart piece by piece and rightly examine it. The steps are all made according to architectural theory. . . . All of it is made and gilded according to the texts and the beautiful description of the Temple written by the famous Villalpando.[25]

Schott's model, as the Uffenbach description affirms, presented a faithful material rendering of Villalpando's temple drawings. The Uffenbachs also described the Schott model as a piece of art, a *Kunststück*, and not as an instrument or a philosophical apparatus like Leon's. Despite their similarities, the reasons Schott and Leon built their models were dramatically different. Just as different were the accounts and practices through which these architects produced them.

Although Solomon's Temple was an extremely important and common point of reference for early modern scholars of all confessional backgrounds, there was no consensus about which texts to use in order to produce a reliable presentation of it. Indeed, the very act of representing a space of justice, wisdom, and reconciliation, a space widely believed to possess universal significance, seemed to perpetuate only confessional strife and difference. Some tried to remedy this problem. Villalpando, for example, influentially combined descriptions of the historical Temple from 2 Chronicles, descriptions of a coming temple by the prophet Ezekiel, and Vitruvius's description of temples in ancient Rome. But Villalpando's overt use of Vitruvius's and Ezekiel's accounts of an imaginary temple made many uncomfortable. Montano faulted Villalpando on exactly these grounds. For his own temple drawings, he used the Mishnah and Middoth, Talmudic sources, which he felt would bring his readers closer to the original Temple than the Old Testament accounts contained in 2 Chronicles or Kings ever could.[26] Unlike Villalpando, Montano privileged Jewish ritual objects, like the candelabra or the vestments of the high priests, over the precise, proportional arrangement of the space itself. In the end, the pope reprimanded Montano for relying too heavily upon Jewish sources. Villalpando's drawings acquired much more fame and esteem, but Montano's drawings continued to be preferred by many Protestants.[27]

The manner in which Semler built and arranged his temple model communicated his interest in moving past the confessional polemics so characteristic of the temple reconstruction projects that had preceded him. The space of the model became a site for displaying, mobilizing, and reconciling a variety of particulars—sources, methods, and aesthetic preferences embraced by Jews, Protestants, Catholics, and "heathens"—into a composite whole. According to Samuel Lange, Semler was an expert in providing "an introduction to the art of observing the whole and all of the parts that fit within it." He taught others how to take "all of the clear philosophical particulars and see them as wheels—well divided wheels that fit somehow into larger whole"; his goal was to help his pupils learn "how to think, to test, to differentiate and to find."[28]

The method Semler used to build his model is best described as eclectic in its aims and orientation. He portrayed himself as exactly the kind of "idea collector" that Johann Christoph Sturm had described in his introduction to the *Philosophia Eclectica*.[29] One's status as an able idea collector was, in turn, defined by his ability to be discriminating while gathering up these ideas. This involved actively seeking out a variety of opinions and being willing to use "any method" in order "to recognize the truth and to distinguish it from the untruth."[30] A practicing eclectic stepped outside his confessional affiliations, obligations, and assumptions and operated in a neutral space. He devised his own coherent strategy for assimilating variety. We can see these strategies in action by considering how Semler built his temple model.

First, and in accordance with his preference for eyewitnesses, Semler said he had confronted the prospect of reconciling firsthand accounts of the Temple contained in Kings and Chronicles. He obviously parsed these texts for valuable information to make visible in his model and struggled with how to handle the peculiar silences, omissions, and contradictions they contained. "It has to be admitted," he wrote,

> that at the beginning, two large challenges presented themselves when building [the Temple]. The first was where the authors let us down; one had to go on building although there was no sufficiently detailed account of the respective next piece of it to be found in any author. The other was where the authors contradict one another; the authors describe the piece to be built next differently, and yet I wanted very much to reconcile them because both of them had a good reputation and were trustworthy.[31]

These trustworthy firsthand accounts sometimes contradicted each other and often left out key pieces of information. Yet Semler claimed to have overcome these challenges. He did this, he said, by only commencing with reconstructing a part of the model when a clear firsthand description of it was available. He also said that eventually, in the act of building, both the connections between various components of the buildings and the causes of the contradictions became visible to him.[32]

Very often, Semler argued, the transmission of ambiguous information about a specific part of the Temple, combined with the lack of attempts among previous commentators to translate this information into something visible or tangible, led to divergent accounts. His model's power stemmed from its ability to fill in a previously ambiguous space with something concrete, thereby helping to generate clearer and more distinct ideas about it. This problem of ambiguity plagued even the most learned and respected of accounts. For example, Semler said he was especially fond of a discussion of the Temple produced by an evangelical minister from Schleßwig named Johannes Lund.[33] Yet despite how much he trusted Lund, Semler said he was forced to stray from his account in one key instance. Lund had argued that the Temple's *spatium intermurale*—the space between the partition and the external walls—had extended sixteen *Ellen* (roughly thirty-two feet) along steps stretching down the Temple Mount. At the same time, the Jewish *Tractat de Mensuris Templi* (chap. 3, §9) indicated that every time a priest went out of the Temple, he had to go "under the Intermurali" through a gate called Teri. When one endeavored to build the Intermurale in a model, Semler explained, its appearance before the eyes (*zeiget es der Augenschein*) showed that it clearly must have been elevated and near the inner courtyard:

> In the model it is evident that the Intermurale was erected close to the inner courtyard because the Israelites wanted to make offerings in a way that involved waiting for each other; this would ensure that when the first sacrifices were offered, other Israelites would be immediately present to succeed them. This could not happen so quickly if the Intermurale had been 16 Ellen deep and every Israelite had to climb 32 steps with the sacrificial animal first.[34]

Through the very act of building and observing the model, Semler became convinced that he had to dispense with this part of Lund's account and rely on an alternate source.

Semler indicated that if Lund were to find out about his decision to stray from his account, he would not be dismayed "because the sight of the model would make the reasons apparent."[35] In other words, Semler was convinced that his model would have been powerful enough to have convinced Lund of the reason for his choice. Lund, in turn, would have been compelled to accept Semler's decision as a valid one. Semler used the model to help him resolve the problem of the Intermurale and, as he did this, he started to see how it could help him eliminate further ambiguities by making their causes visible.

The model also made manifest several forms of expertise at the same time. Every account of the Temple contained mistakes but was useful, Semler argued. He tried to illustrate what he meant by describing the circumstances that compelled him to begin building his model. In early August 1716, he wrote, a rabbi from Prague had arrived in Halle with a temple model made out of white wax. Roughly three feet long and wide, it was a striking object that many, including Semler, went out of their way to see. The model made it clear that the rabbi knew much about the ancient Temple, but Semler insisted it was also full of mistakes. He said it was not built to scale, nor did it show that its builder was experienced in the mechanical arts or in architecture. It contained spires and slanted rooftops like Christian churches, which buildings in antiquity simply did not have. The structure's two main pillars, Jachin and Boas, were supposed to be freestanding, but in the wax model they were built into the walls. There was also no upper room in the model, but the real Temple had one.

Despite these "mistakes," Semler said he valued the rabbi's contribution and endeavored to include him in a new model-building project. Semler wrote that they had successfully collaborated to complete the new wooden model's foundation before the rabbi returned to Prague. At this point, Semler said he began to rely upon sources the rabbi also trusted, especially Talmudic authors who had special expertise because they had written in close proximity to the Temple while it was standing.[36] Semler noted that L'Empereur's translation of sections of the Talmud into Latin had made it easier to access these texts. He also appraised Flavius Josephus's *Antiquitatibus Judaicis* and *de bello Judaico*, explaining that it contained "a few very good" pieces of information about the Temple but that these pieces were "not complete or enough." He consulted Moses Maimonides's *Tractat de domo electa* (translated by Ludovicus de Compiegne de Veil) and the descriptions offered by Jacob Jehuda Leon in *De templo Hierosolymitano*. He then consulted the English theologian and Hebraist John Lightfoot's temple ground plan, dismissing other elements of his presentation.

All of these contributions were useful, Semler wrote, but still "not enough to build a complete model."[37] This required even more expertise, including training in the mechanical arts and architecture, and the construction of a ground plan upon which to anchor and reconcile all the physical elements of the space. While working with the rabbi, Semler created a precise, ichnographical and skenographical footprint for his model (fig. 9). In this sketch, which involved the use of a compass, we see especially the thickness of the model's walls and the precise arrangement of ritual objects inside it. Semler noted that he actually affixed the original ground plan to the square wooden board that served as the model's foundation after completing it. He further elaborated on the significance of

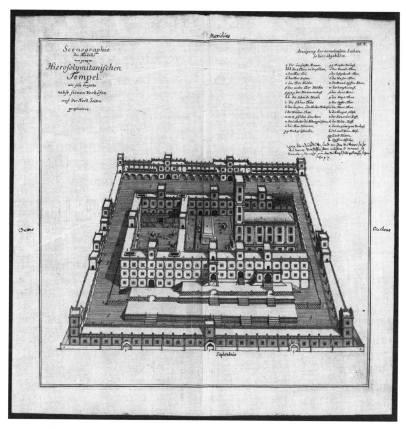

9 Halle's model of Solomon's Temple as a *skenographia* from Christoph Semler, *Der Tempel Salomonis* [Halle: Waysenhaus, 1718], engraving. Credit: Niedersächsische Staats- und Universitätsbibliothek, Göttingen [8 H AS 1, 3016].

the ground plan in a lesson he wrote for his *Realschule* in 1707–1708, describing it as the best kind of "perspective drawing" because it was capable of presenting "all sides of a building at once."[38]

Writing from Halle around the same time Semler was building the temple model, Christian Wolff also defined the ground plan as a *"Maas"* that reconciled various heights, widths, and depths with each other "to establish a certain ideally simple, proportional relationship" and explained that these relationships held the key to a universal mathematics capable of stirring the soul.[39] According to Meier, in this period Semler was also offering instruction in "pure mathematics (*Mathesi pura*)" in his home and teaching his pupils how to think scientifically (*zu der scientischen Art zu denken*) by demonstrating clauses from Wolff's *Beginner's Guide.*[40] Semler was clearly interested in geometrical projection techniques and the Temple's proportions; he alluded to Lund's interest in determining "each and every Elle and half-Elle" and noted that ascertaining the Temple's precise measurements was "also a goal of our project, so that one would be more certain in the construction of every part of this small Temple."[41]

Semler described his interest in using his model to promote consensus by convincing those who observed it that it was the most reliable replica around. He deployed techniques he hoped might convince those who visited his model to accept it as correct—basically walking observers from a variety of social and confessional backgrounds through the building process. He tried to help them understand the reasons he had elected to make the choices he did in conjunction with the meaning of each piece. In many ways, he was inviting them to participate in the reconstruction of his model. And while I think he wanted to generate certainty about the original Temple through his model, he recognized that the best he could do was to inspire others to accept his model's provisional status and to apply their own expertise—to be open to collaboration with other experts—toward perfecting it further.

The model emerges in the pages of Semler's viewing guide as a thing that perfects and is perfectible. It even had removable parts. It was built in such a way that portions of the roof and walls could be added and removed as needed. One could examine the Temple's eastern facade as it looked from the outside and take off a wall and look inside the main building (figs. 10 and 11). While peering inside the structure, one saw that the main Temple and its tower were divided up differently than they appeared to be divided from the outside. From the outside it seemed that the tower had four stories. From the inside, it was clear the tower actually had three stories, but the first of these stories was unusually extended. One saw with one's own eyes that the Temple could easily be construed as having three, four, or even five stories.

The Temple tower, like the Intermurale, was an ambiguous space; this ambiguity generated controversy. For Montano, the Temple's eastern facade had five stories (fig. 12). Villalpando, on the other hand, had presented a three-story eastern facade (fig. 13). Semler's model contained both of these variations. By building his model the way he did, Semler illustrated how at least two authorities offering conflicting opinions on this matter were in some way correct. He made it clear that there was a middle way between Villalpando's and Montano's expertise.

Semler endeavored to reconcile competing but trustworthy claims by making the conflicting firsthand accounts visible in his model. He wanted

10 Halle's model of Solomon's Temple, main building with tower and external walls, from Christoph Semler, *Der Tempel Salomonis* [Halle: Waysenhaus, 1718], engraving. Credit: Niedersächsische Staats- und Universitätsbibliothek, Göttingen [8 H AS 1, 3016].

observers of his temple model to acquire better ideas about how to initiate new technologies of reconciliation in the world. This subtext is especially poignant in the sections of Semler's viewing manual where he offered an account of Yom Kippur and in a long section on the tabernacle, the portable sanctuary used by the ancient Israelites while in captivity. The ancient Israelites funded the construction of this structure out of their own pockets and their own free will so that God would have a place to dwell among them while he helped them with the "reconciliation of their souls," Semler wrote. The money that the "children of Israel" paid to build and to maintain the tabernacle was called the "money of reconciliation" (*Geld der Versöhnung*), and it was laid on the altar during worship

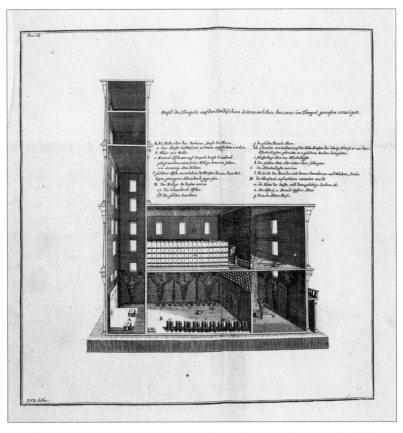

11 Halle's model of Solomon's Temple, main building with tower and no external walls, from
Christoph Semler, *Der Tempel Salomonis* [Halle: Waysenhaus, 1718], engraving. Credit:
Niedersächsische Staats- und Universitätsbibliothek, Göttingen [8 H AS 1, 3016].

services.[42] Semler's temple model and manual encouraged observers to
understand these donations as acts of friendship, love, and philanthropy.

Media for the Inner Eye

Writing to the eight-year-old Prussian crown prince Friedrich II (later
Friedrich the Great) in August 1720, Francke described how certain kinds
of models participated in initiating processes of awakening and illumination. His letter accompanied a model of the city of Jerusalem that he
sent to the prince in Berlin. In the letter, he wrote that he hoped the
model would help awaken the boy "to desire and love the holy word." He

12 Main building and tower of Solomon's Temple, from Benito Arias Montano, *Antiquitatum Judaicarum*, book 9 [Leiden, 1593], engraving. Credit: Herzog August Bibliothek Wolfenbüttel [A: 58.4 Hist].

explained that viewing this "form of Jerusalem" would make the process of constructing the prince's character more pleasant, since "the frequent observation of it and the comparison of it with the Holy Scriptures . . . can yield all kinds of benefits, so that the word of God is approached with more and more desire."[43]

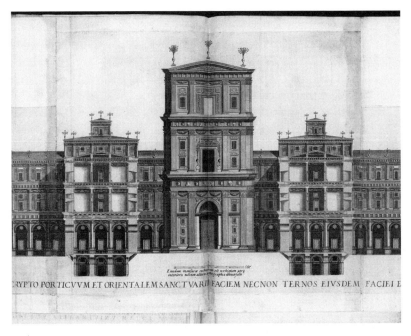

13 Main building and tower of Solomon's Temple, from J. Villalpando and J. de Prado, *Ezechielem Explanationes* [Rome, 1596–1605], engraving. Credit: Niedersächsische Staats- und Universitätsbibliothek, Göttingen [2 TH BIB 816 /10:2].

Francke frequently advocated the systematic observation of virtuous and beautiful examples or models. In pedagogical writings such as *Short and Simple Lessons*, he advised that in the schools virtues and vices "should be painted in living colors before children in order to strengthen the good that has been initiated in them." *Vormalen*, or vividly depicting vice and virtue, he continued, "has been determined by reasonable heathens as a good way of awakening virtue and getting people to turn away from vice."[44] He pointed to Theophrastus's *The Characters* as offering the perfect example of how to do this. Several Christian authors, such as Joseph Hall and Johann Heinrich Boecler, he noted, had provided a summary of signs of virtuous characters derived from Theophrastus's typology from which lessons inspiring the appropriation of these signs could be derived.[45] Boecler was a professor of history in Strasbourg who created a typology of virtuous Roman emperors based on the work of the Roman historian Vellejus Paterculus. Hall, an Anglican bishop, read Theophrastus's *Characters* as a work of moral philosophy, which he said inspired him to publish a two-volume collection of character sketches in 1608. The collection became so popular that it would eventually spark an entirely new yet derivative literary genre.[46]

Circulated with posthumous commentary by Isaac Casaubon in 1688, the *Characters of Theophrastos* presented its readers with a collection of signifiers of virtuous characters. Despite the promise in the introduction of examples of both virtue and vice, Theophrastus's characters were all archetypes of vice, whose conduct and other observable signifiers provided unencumbered access to the state of their souls.[47] Christian authors who endeavored to follow in Theophrastus's footsteps were concerned with creating signifiers of virtuous archetypes as well, in the interest of inviting emulation. Joseph Hall described one who is wise as someone who is "eager to learn or to recognize everything but especially to know his strengths and his weaknesses."[48] His eyes "never stay together, instead one stays at home and gazes upon himself and the other is directed towards the world . . . and whatever new things there are to see in it."[49] The eyes of a properly developed Christian are also one of his most clearly recognizable signs. They are "illuminated" and rest upon the uncertain in a way that directly affects his senses. "He speaks with God in prayer" and "when he comes near to God puts on a purple robe."[50]

Throughout the *Short and Simple Lessons*, Francke also noted the importance of the French church historian and educator Claude Fleury's *Traité du choix et de la méthode des études* (1686). Fleury was part of a circle of pedagogues in France called the Petit Concil, which became famous for numerous essays about how best to cultivate virtuous rulers; the circle also included Jean de la Bruyèr, a teacher of Louis XIV's cousin, and François Fénelon.[51] Insisting upon the importance of the Petit Concil's advice, Francke encouraged the instructors of the Orphanage schools to create character typologies using famous biblical personalities. He told them to strive to use "living and exemplary" presentations of archetypes, such as Joseph, whenever possible, to "awaken a love of virtue in children so that they wish nothing more than to be similar in every way to these figures."[52] This, of course, is precisely what Fleury had suggested in his *Customs of the Israelites* (*Moeurs des Israelites*, 1681).

The picture accompanying the title page depicts two men gazing at two Israelites from a slight distance (fig. 14). One is writing an inscription of what he sees in front of him, while the other is holding a large canvas on which he seems to be drawing a likeness of the two figures. In addition to intense work on the self and the observation of virtuous archetypical character types, a signature feature of pedagogical practice in the Orphanage became the reproduction and emulation of these kinds of

Sitten der Ifraeliten.

14 Example of "Vormalen," from Claude Fleury, *Die Sitten der Israeliten*. [Hannover, 1709], engraving. Credit: Archiv und Bibliothek der Franckesche Stiftungen [BFSt: 64 F 32].

models. In his *Admonitory Lectures* (*Lectiones Paraenetica*), Francke noted that the scriptures were full of models of sanctified lives, offering a mirror that forced us "to confront our ugly appearance."[53]

As Francke had told the crown prince when he sent him his own model, the benefits of observing beautiful or morally upright models triggered special kinds of pious desires and a special kind of prudent knowledge or sagacity called *Klugheit*. There were true and false forms of *Klugheit*, he argued. Each one

rests upon two main pillars, namely knowledge [*Wissenschaft*] or understanding [*Erkenntnis*] and experience [*Erfahrung*]; either pillar can be misused on its own but together they can both be used correctly. And *the true Klugheit is nothing other than the eye in people*, through which one sees what serves the best and protects oneself from damage.[54]

A good model, in other words, triggered or helped generate *Klugheit*, which Francke defined here as "the eye *in* people." Note his emphasis on how this inner eye rests on the two pillars—one associated with the mind (knowledge, understanding) and one with the body (experience and the senses). In light of the ongoing discussions in Halle about how to create new theories of emotions consistent with a mechanical worldview and to reconcile empiricist and rationalist approaches to knowing the world, Francke's interest in *Klugheit* as the inner eye makes a good deal of sense. He wanted the members of his community to associate the inner eye with forms of cognition that hinged on the observation of exemplary and especially perfected real things.

If we had a chance to ask him about it, I think Francke would have insisted that the observational techniques his teachers deployed in the Orphanage intentionally triggered certain desires or inspirations that improved the learning process (made it fun)—*and* helped young people learn how to observe in a methodical, orderly way. This was part of his effort to set his community apart from those of religious enthusiasts or *Schwärmer*. Critics of enthusiasm, such as Henry More and Meric Casaubon, had long argued that these individuals suffered from an excess of "ecstatical passion" that reinforced "*illusions of the imagination*."[55] Their visions and ecstasies meant they could not see clearly or methodically. They were not selective about the kinds of models they chose to emulate, nor did they endeavor to carefully manage their desires and passions. Most important, they did not know how to put things in perspective; therefore, their claims of divination posed dangerous and unpredictable threats to the social order. Confronted with the evidence of the staggering variety of observational exercises undertaken in the Orphanage, anyone who might be tempted to describe those involved with it as *Schwärmer* would have been challenged to rethink these kinds of assumptions.

As young people observed Semler's temple model, they were given opportunities to work on improving their ability to sharpen their memories of all its components, including relationships of individual parts to each other and to the respective whole.[56] Teachers tried to help them better understand the model's significance by reading short sayings copied from sacred books that were posted on a screen placed next to the model. They also encouraged their pupils to move around it, to get their bodies involved in the process of observing it—standing at different angles and observing it from different vantage points. And even when they could not stand directly in front of it, Semler's manual served as a tool they could use to virtually witness the model and even to relive the experience of having observed it in person, thereby continuing to engage the inner eye as frequently as possible.[57]

Putting Models in Perspective

After the 1721 curriculum revision (see chapter 3), young people's encounters with Semler's temple model served as a starting point for all of the other exercises they undertook during their periods of recreation. Freyer noted in his description of the new methods being deployed in the *Pädagogium* that pupils moved according to a "previously prescribed order" so that their "bodies and characters could undergo a positive transformation."[58] The exercises they undertook after their warm-up observational exercises in "physics and the Bible" involved models as well—including making their own models of things and learning how to put models in perspective through a series of exercises they undertook in the mechanical disciplines. Approximately thirty pupils at a time turned wood at individual turning stations; each one was expected to acquire all the materials he needed and to "decide for himself what he wanted to make."[59] The goal of these exercises was not for the students "to make something nice and finished for them to take home with them," Freyer noted, but rather "the knowledge [*Wissenschaft*] itself and the motion."[60] As they improved, their instructors would ask them "to make models from different materials and give them to the teacher so that they could serve not only as examples [of the things they had completed] but also as options or choices" (fig. 15).[61]

In addition to making models out of wood, Freyer explained that pupils could also make models out of cardboard in a small workshop (*Papp*

15 *Modell des Waisenhauses mit Blick aus einen imaginären Raum auf das Hauptgebäude der
Franckesche Stiftungen* (model of the Halle Orphanage with a view of the main building from
an 'imaginary room' in the ensemble), from Johann Friedrich Penther, *Einer Ausführlichen
Anleitung zur bürgerlichen Baukunst*, part 2 [Augsburg: Pfeffel, 1745], engraving. Credit:
Archiv und Bibliothek der Franckesche Stiftungen [BFSt: 132 A 8a].

Fabric): Specifically they made boxes, little cabinets and chests, apothecary containers, stereometric bodies of different geometrical figures, and other "useful things" that they would need to exercise in practical mathematics.[62] Before learning how to use the machines associated with these sciences—such as the microscope or a camera obscura—pupils first had to make models of these machines.[63] Only after they had done this were they allowed to grind glass lenses for use in "English microscopes," burning mirrors, small perspectival devices, telescopes, and camera obscuras, magic lanterns, reading glasses, and machines for creating ground plans for buildings and cities.[64] The best pupils were taught how to use the machines themselves to create different optical effects.

To help with lessons in astronomy in the *Pädagogium*, Francke commissioned Semler to build the two giant models of the universe mentioned by Samuel Lange and Georg Meier in the excerpt cited earlier, the heliocentric and geocentric heavenly spheres. The models presented the systems of both Tycho Brahe and Copernicus. Lange reported that "both are ordered in such a way that the precise movement of the whole universe and individual pieces, including the moons of the planets, can be guided with a single twist so that they agree with real time, eclipses and all other movements."[65] These models took Semler a total of three years to build and provided observers with outstanding examples of two competing models for understanding the universe.[66]

As with his temple model, Semler developed elaborate scripts to help student teachers present each one. Teachers were to explain how the spheres were made, including the materials they were made of. Semler insisted that each model, when demonstrated and studied together, improved the observer's ability to understand the structure of the heavens. This was largely because, he noted, the ability to observe different models of the universe at the same time was highly unusual, allowing young people to better appreciate the similarities and differences between the two systems.[67] Semler stressed his dueling models' potential to help individuals learn to see the beauty of things according to their proportional relationships with each other. "The heavens are the most beautiful parts of the world," Semler wrote, and this world "is like a mirror wherein one sees God all at once."[68] Since "the human eye cannot take in the entire world and its broad reaching heavens at once," Semler explained, "people have invented a few useful machines that are current and present imitations of the heavens and its spheres."[69]

Semler's models of the Tychonic and Copernican systems also allowed observers to practice seeing things from different points of view and to understand relationships of scale. As they observed the models, pupils were to be told that "for those of us who live on Earth, the heavens appear as a large, round ball, which surround us halfway with their concavity."[70] They were also told that through these models "the convexity of the heavens can be presented, as if we were extended outside of and above the heavens."[71] While demonstrating the system of Tycho Brahe, the teacher did not say whether or not it was right or wrong. He noted instead, for example, that "in this model, the earth is presented much too largely. Compared to the heavens it should only be the size of a poppy seed or smaller."[72]

Semler designed other machines used in conjunction with these models that were supposed to enhance a person's abilities to observe by helping them suspend themselves "outside of and above the heavens." His comet machine displayed "all the comets that have been seen at some point and can be shown according to a specific hypothesis." His *lucerna astronomica* "show[ed] solar eclipses, the phases of the moon and present[ed] all asteroids in a very lifelike manner." Additionally, Semler built an instrument he said allowed young people to practice seeing the earth as a non–earth dweller might. "Using the small instrument Lit: E, *Astronomia Selenitarum* will be presented," Semler noted, "or how we would see the world if we were on the moon." Since the moon "turns against the earth all the time," he continued, "if there were people who lived on the side of the moon that faces the earth, they would constantly see the earth." Despite the relative frequency with which they would see the earth depending on their position of residence on it, "it would appear to their eyes four times larger than the moon appears to us."[73]

Observers on earth, while urged to suspend their formulation of reality and consider the perspective of the moon dweller, could also come to better appreciate the fact that the moon sometimes appeared near and sometimes far to their eyes and to know why this was the case.[74] "The sun is sometimes very close to them and sometimes very far away," he noted. "This means that the sun will sometimes appear to their eyes to be very big and sometimes it will appear very small." In all of these instances, a discussion about what "moon dwellers" see was not about whether or not these people really existed but about using their point of view to help pupils learn how to appreciate and understand other perspectives, to empathize with these alternative ways of seeing the world by putting oneself in someone else's shoes.

In addition to making and observing a wide variety of models in the Orphanage, student teachers and pupils also had the opportunity to practice drawing them. In 1717, while Semler's temple model was still under construction, the Orphanage published a drawing guide offering an introduction to *Graphice*. The guide stressed the ability of projective instruments like the camera obscura or the magic lantern to help pupils learn how to become proficient in the art of perspectival drawing.[75] Even after the curriculum revision, student teachers were using this guide in the Orphanage schools. In the *Pädagogium*, Freyer explained:

All begin together from the very beginning and learn first how to draw with pencil on paper and then with chalk on a board; and to be sure, they sketch according to copperplate engravings that are placed in front of them as models, where they can proceed from the easiest to the more difficult by first depicting geometrical lines and figures and then natural and artificial things in the easiest and most fundamental way, one after the other.

. . . After this they move toward a short introduction to light and shadow and then toward coloring and painting after nature. Whereby they either copy something already drawn or even something more difficult, like emblems or strange scenes or coats of arms or even perspectival things can be used: until finally an entire landscape or history could come out of it.[76]

Freyer noted that when a student was "still unable to do perspective," then the instructor would "give him a short introduction without many rules" and "be satisfied with especially simple or poor drawings."[77]

When the Swiss natural philosopher, physician, and member of the British Royal Society Johann Jacob Scheuchzer mentioned the Orphanage's temple model in his *Physica Sacra*, he was careful to note that it was laid out before the eyes "clear, distinct and bright" and that one could observe it "in perspective." By the 1730s, one of the temple models stood in the middle of the Orphanage's museum, flanked on either side by models of Jerusalem and the Holy Land, the tabernacle, and Semler's Tychonic and Copernican models of the solar system.[78] That one could view the temple model in perspective meant one could view it in context, alongside models that helped clarify its geographical location and relationship to other noteworthy buildings—such as the tabernacle. It also meant that one could observe it and, with the help of Semler's handbook, cultivate an awareness of the variety of sources he had consulted to build the replica and the reasons he chose to depict certain components the way he did—particularly when he had competing perspectives or options to choose from. One could exercise the memory by repeatedly observing the temple model's various components and their relationships to one another to imprint an image of the model in the mind; one accomplished this best by walking around the model and observing it from various vantage points.

Finally, the model was an ideal tool for triggering intuitions, pious desires, and the inner eye.

Constructing the temple model was a long process that required some understanding of practical mathematics, including projective geometry. It represented the organization's commitment to an eclectic biblico-historical worldview pursued via a sacred or mosaic physics. Even more important, publicizing the possession of such a large model provided Orphanage administrators with a special tool to clarify what set their community apart from others. Whereas other radical or "enthusiastic" religious groups, including Anabaptists, Moravians, and Zionists, also made sacred architecture central to the construction of their new communities, they relied mainly on descriptions of a "heavenly Jerusalem" found in the book of Revelation, where there is actually no temple mentioned at all.[79] The Orphanage, on the other hand, was supposed to be a different kind of community: showplace, temple, "universal seminar," and "prophet school."[80] As the Lutheran theologian Johann Michael Dilherr had explained in his book *Prophet School* (1662), these were places where young prophets in training came to learn from resident visionaries, or "truth-sayers," who spent most of their time helping their pupils acquire the same heightened abilities to observe.[81] They were not fanatics; they had simply cultivated special ways of observing and apprehending things that most others could not see.

FIVE

An *Observator* and His Instrument

The Halle Orphanage became the headquarters of scientific missions to Russia, southern India, and North America. To expand its reach, the organization relied on affiliated student teachers and pupils to apply their expertise in practical mathematics and experimental physics as skilled observers. This was a key part of what it meant to reimagine the school as a place for collaborative research; indeed, a major reason Leibniz and Tschirnhaus were so interested in the Orphanage was that they saw it as a kind of academy that would help create a network of highly trained scientific observers. Talented young men were to emerge who would be tolerant of myriad points of view and committed to solving real-world problems. Leibniz was interested in training a particular kind of observer—he used the term *Observator*—who would be "not only a mathematician, . . . but also a student of nature, who observed plants, animals, minerals and other *naturalia* and *artificialia loci*, because this all goes together and can be done at the same cost."[1] He noted that it would be important for skilled *Observators* to be technologically savvy and able to travel with special instruments, especially odometers, compasses, quadrants, pendulum clocks, levels, microscopes, barometers, magnets, and magnetic globes that would help them apprehend things in the world most people missed entirely.[2]

In this chapter I tell the story of a traveling *Observator* affiliated with the Halle Orphanage named Christoph Eberhard. Eberhard's travels took him to Denmark, Holland, England, and across Russia, where he spent several years in the inner circle of Peter the Great while serving as the house chaplain of one of Peter's most trusted advisors, General Adam von Weyde (Adam Adamovic) of Moscow. Eberhard witnessed the Great Northern War (1700–1721), spent time in an ironworks—where he led Pietist group meetings, or conventicles—and became involved in efforts to found a facility modeled after Halle's Orphanage in Tobolsk, Siberia. He played a crucial role in generating a buzz about the new educational ensemble in Russia, particularly among the tsar's advisors, who sent their children to be educated in Halle and even visited themselves. He later became known for the *Instrumentum inclinatorium* (hereafter referred to as the inclinator) he built with the help of his friend Christoph Semler in Halle. This was the special instrument for measuring geomagnetic phenomena that Francke showed King Friedrich Wilhelm I during his visit to the Orphanage in 1720. Just two years earlier, Eberhard had taken this instrument to London, where he had told the members of a parliamentary commission known as the Board of Longitude that it could be used to measure longitude reliably and across great distances.[3]

I believe we must see Eberhard as a product of the collaborative relationship forged between the Orphanage, the University of Halle, and the Berlin Academy in their earliest phases. His activities contributed to growing international interest in the Orphanage as a scientific community. Eberhard spread word of the Orphanage as a showplace for a new culture of public science harnessed to philanthropy and worked to promote the "academic reputation" of the University of Halle as well.[4] This chapter discusses his travels and efforts to make the systematic observation of *magnetic inclination* the key to uncovering the universal cause of all geomagnetic effects. The final section untangles what is undoubtedly the most perplexing part of Eberhard's story: he eventually hid his connections to the Orphanage and became estranged from the very community that had sent him abroad as a scientific observer in the first place.

The reasons for this estrangement were linked to the mounting tensions surrounding how to practice scientific eclecticism and efforts to institutionalize mosaic physics in Halle discussed in chapter 3. In short, Eberhard became caught up in the same set of debates about the power of the theology faculty that contributed to Christian Wolff's expulsion from the city. His efforts to create his own magnetic theory ultimately were not in step with efforts to sanction the production of explanatory frameworks or syntheses among those affiliated with the Orphanage. In the end, the organization rejected Eberhard, while continuing to associate itself with the inclinator. Eberhard's story is worth recovering in part because it allows us to see how Orphanage *Observators* promoted and extended the reputation of the community abroad; it also helps us begin to appreciate why the Orphanage's reputation as a scientific community began to suffer in the 1720s as it started to reject certain individuals and perspectives entirely.

On the Move

Born the son of a preacher in Eisenach, Eberhard came to Halle to study in 1703; he was twenty-eight years old, a "poor student" who supported himself by working as a teacher in the Orphanage. In 1704, Francke sent him to Copenhagen, then to London, to rally support for his plans, which included sending theology students on scientific missions to the Danish colony of Tranquebar in southern India. While in Denmark, Eberhard served as a private tutor (along with Johann Otto Glüsing) and led Pietist conventicles composed of a combination of university students, merchants, and artisans; he also connected with Heinrich Wilhelm Ludolf, already a trusted agent of the organization and a close friend of the tsar.[5] Eberhard's role abroad initially involved asking people for money and helping potential supporters understand the aims of the Orphanage. Writing to Francke from Copenhagen on March 20, 1706, Eberhard told him that upon hearing about Halle's Orphanage, a General Major Husman "had resolved to build an orphanage in Christiana after the model of the one in Halle."[6] Inconveniently for Eberhard, conventicles were outlawed in Copenhagen beginning in October 1706 when the then court chaplain, Franz Julius Lütkens, convinced the king that attendees were reading "anti-church" writings; Eberhard and his friend Glüsing were forced to leave Copenhagen shortly thereafter.[7]

Letters exchanged between Francke, Eberhard, and other administrators from roughly 1707 to 1711 indicate Eberhard was on his way to becoming one of the Orphanage's most trusted agents abroad.[8] In 1710, Francke sent him to London, where he met another agent for the organization, Johann Tribbechow, then serving as a court chaplain for Prince George of Denmark.[9] Unlike Eberhard, Tribbechow had become a bit of an embarrassment to his supporters in Halle: he had had an affair with a merchant's daughter whom he later claimed had given him a philtrum, or love potion. Eberhard helped smooth things over, writing directly to Francke about Tribbechow's "illness" and suggesting he allow the pastor to quietly slip away and avoid further damage to Halle's reputation.[10]

This was a crucial intervention because, at the time, Francke was still in touch with the Society for the Promotion of Christian Knowledge. In 1698, he had launched a campaign to promote the Orphanage that was especially well received by the founders of the society and sent two student teachers, Jacob B. Wigers and Johann C. Mehder, with explicit instructions to attend an SPCK meeting. They must have made a good impression, because shortly after their visit, SPCK leader John Chamberlayne told Francke:

I am to acquaint you that they [the Society] desire to maintain a Frequent intelligence with you, both to be informed of the great things you have already done in Germany towards the promoting God's Glory and spreading of Christian knowledge etc., and also to communicate to you what Measures they take in the carrying on of the same Designs etc.[11]

A few years later, the SPCK produced propaganda stressing the innovative qualities of the Orphanage; it featured the main building as the new symbol of Halle Pietism (*Pietatis Hallensis*) (fig. 16). The SPCK also translated Francke's *Pietas Hallensis or the marvelous Footsteps of Divine Providence*, in which he told the story of the Orphanage's founding. He continued to worry about the organization's reputation in London, including its ability to attract benefactors outside German territories—thus the need for Eberhard's help.

About a year after Francke sent him to London, Eberhard received a new assignment. He was to go to Russia to serve as household minister for General Adamovic, Count von Weyde, in Moscow—another huge opportunity for him. Von Weyde was one of a handful of close advisors that Tsar Peter I had chosen to travel with him during his first Grand Embassy of Western Europe in 1697–98. Along with Jacob Daniel Bruce and the Norwegian shipbuilder Niels Olsen (a.k.a. Cornelius) Cruys, whom the tsar had recruited in Amsterdam and made one of his admirals, von Weyde toured shipyards, artillery factories, and much more during this trip. Upon returning to Russia, von Weyde became immediately involved in a series of conflicts between Russia and Sweden known as the Great Northern War.[12] He was taken prisoner by the Swedes in the Battle of Narva in November 1700 and imprisoned for eleven years, after which he was released with the tsar's help. Prior to these experiences, von Weyde met Heinrich W. Ludolf in Moscow and learned about the Orphanage.[13] Upon his release from Swedish captivity in 1711, he specifically requested a minister for his household from Halle.

Von Weyde's request reflected the growing interest in Halle's Orphanage in Russia and marked a great opportunity to grow the organization's long-distance network further. Indeed, when Francke wrote to Leibniz in 1698 to tell him he had read the *Novissima Sinica* and to let him know

They that wait upon the Lord

Shall renew their strength

that "the field in Russia" had already been opened, he meant in part that there were already emissaries from the Orphanage operating there. Upon arriving in Moscow, Eberhard was met by two of these emissaries, Peter Müller and Justus Samuel Scharschmidt. Scharschmidt had been hand picked by both Francke and Spener to forge new connections with Protestant groups in Russia.[14] Müller was from a German family then living in Ugodka, near Moscow, who owned an ironworks and had connections with the tsar.[15] He had met Eberhard while studying theology in Halle from 1700 to 1704—just before Eberhard left on his first big assignments in Copenhagen and London.[16]

While in Russia, Eberhard increasingly found himself engaging with

Russian literati at the center of the tsar's own scientific and educational reform initiatives, which included creating a scientific academy and a museum (*Kunstkamera*) as an instrument for public education.[17] He must have met von Weyde's friend Jacob Bruce, for example, who had ties to Halle because he had hired the son of one of Francke's oldest friends, Gottfried Vockerodt, as a tutor for his nephew. After meeting Jacob Bruce in Dresden in 1711, the younger Vockerodt agreed, against his father's wishes, to travel to Moscow, where he taught geography, genealogy, and heraldry for the Bruce family.[18] His father managed to convince both Francke and Eberhard to help him persuade his son to come back home to Saxony or Brandenburg-Prussia, which would have meant spending time in the Bruce household.[19] Bruce, like von Weyde, had accompanied the tsar on his first Grand Embassy tour and was interested in navigation, astronomy, and geography. He was also involved in encouraging Peter I to undertake geographical expeditions to his northern territories and helped start Russia's first mining collegium as well as a school to train mining engineers in Moscow.[20]

From roughly 1711 to 1716, Eberhard accompanied von Weyde on a variety of military campaigns and diplomatic trips throughout the Russian Empire. While living briefly in a military encampment with von Weyde in 1711, he met the Russian theologian Theophan Prokopovic, who was also a professor of poetry at an academy in Kiev, where he taught physics, arithmetic, and geometry.[21] Prokopovic had studied at the Jesuit college in Rome and was fundamentally opposed to the scholastic-inspired curriculum of the order's schools that he had witnessed.[22] He wanted to start new schools in Russia and discussed his plans for educational reform with Eberhard.

Eventually Eberhard got involved in attempts to start a new school and orphanage modeled after Halle's in Tobolsk.[23] These attempts attest to the importance of the Halle Orphanage's ability to serve as a viable institutional model for such efforts. Eberhard recorded his impressions of it and collected letters detailing the "condition of several Swedish prisoners of war also imprisoned in Russia" while living among them there. They were later published as a series of reports, under the pseudonym Alethophilus (Friend of Truth) beginning in 1718.[24] In the first report, Eberhard posed a series of rhetorical questions that suggest he was thinking about the mysteries of the magnet during this period:

Do we have to be simply content with natural things even if we are not able to uncover all the circumstances surrounding them? For example, if our reason should say, why is the magnet so strong in this location and in another location so weak, it [our reason] might become somewhat alarmed. Notwithstanding the fact that it cannot provide an explanation for this, it then becomes common knowledge that a magnet is something that varies in different ways in different locations. But that our reason will simply be content, well there is really no ground for this.[25]

Eberhard communicated his interest in developing explanations for inexplicable phenomena in this passage.[26] It also suggests he was involved in systematically observing magnetic phenomena at this moment, a practice he may have already begun while spending time at the Müller family ironworks, where he also would have had many opportunities to observe magnets' mysterious behaviors.[27]

By 1718, Eberhard had devised his own *Attempt at a Magnetic Theory of the Earth* (*Versuch einer Magnetische Theorie*). In the introduction, he told the story of how two years earlier he had traveled to Danzig with General von Weyde to meet the tsar on official business.[28] He remembered that Tsar Peter had asked him then if he thought it would ever be possible to measure longitude reliably. At this moment, Eberhard became inspired and determined to resolve this pressing problem once and for all.[29] He then left von Weyde's service and returned to Halle's Orphanage, where he conferred with his friends there about building a new observational instrument that could accomplish this goal.

Observing Inclination

By publishing his *Attempt at a Magnetic Theory of the Earth*, Eberhard hoped to prove to the tsar that it was possible to find a way to measure longitude reliably, thereby enhancing his own reputation and that of the institutions in Halle with which he was affiliated. For him the solution hinged on making an original contribution to a conversation that had been ongoing for centuries: namely, how to understand the underlying cause of the earth's geomagnetism. He knew that natural philosophers had long been accustomed to viewing the magnet as a kind of model of cosmic forces at work in the universe and were fascinated by magnets' remarkable abilities to generate motion in certain kinds of materials. He also knew that still no one had managed to explain exactly *why* magnets behaved in the ways that they did. He knew this because of his own experiences with magnets and because he had engaged with a collection of theories about magnets and the reasons for their behaviors that had emerged by the end of the seventeenth century—a "magnetical" philosophy.[30]

The practice of magnetic philosophy brought together natural philosophers, seamen, and those who crafted magnetic instruments, especially compasses.[31] In effect, it became a productive middle ground, wherein scholars and craftsmen, even proponents of both heliocentric and geocentric systems, were able to converse with each other.[32] It came to be closely linked to the experiments of the English physician William Gilbert, who, in 1600, published a treatise in which he posited that the earth was a gigantic magnet.[33] Gilbert's efforts appealed to many people, including Johannes Kepler, who extended Gilbert's claims by describing the sun as a magnet that propelled the planets to move around it.

Eberhard knew that Jesuits in particular had made original contributions to magnetic philosophy, cosmology, and geography. By the 1640s Jesuit missionaries in China and India had begun recording observations of geomagnetic phenomena with universal measuring devices and were sending their accounts back to Rome.[34] As missionaries sent back their observations, the Jesuit polymath Athanasius Kircher assembled them. He described this as a key part of his *Consilium Geographicum*, or great geographical plan.[35] Through its implementation, the Society of Jesus gathered enough empirical evidence to prove that the phenomenon that compasses measured, magnetic *declination*, was not always the same when measured repeatedly in a single location, say Berlin, over time.[36] Eberhard was especially interested in this assertion and in the observational data—reports of magnetic *declination* and *inclination* measured at various locations—upon which such an assertion was based.

I will give a detailed account of *inclination* below, but first a few words about *declination*. Compasses measure declination, that is, the angle between the northern end of a compass needle and the degree to which the needle "declines" from this point in a particular location.[37] Members of the period's scientific organizations, including the Society of Jesus, increasingly found themselves asking this: if one took a consistent series of measurements with a compass in the same spot every day, why was it that the degree to which the needle declined was rarely if ever the same? The more they wrestled with this conundrum, the more they saw the compass as a scientific instrument capable of generating piles of observations that might solve the puzzle one day.

Some individuals, such as the early British Royal Society fellow Edmund Halley, began to devise their own solution to the puzzle using the observations of declination already in circulation. Halley's theory became very important to Eberhard, so it is worth mentioning here. He posited that declination could not be consistently measured in one location because there was a magnetic globe inside the earth "suspended inside a fluid medium" and slowly rotating relative to the outer crust.[38] As Patricia

Fara has explained, Halley believed that both the globe and the earth's crust each had north and south poles and that "the perpetual independent movement of these four poles would cause constantly fluctuating magnetic effects."[39] By 1700, Halley had developed his famous "Atlantic chart," in which he explored ways of representing magnetic declination through the use of isogonic or isorhythmic lines.[40] The chart was similar to the world map showing prevailing wind patterns that he developed in 1686, which allowed observers to "grasp the 'connections of things'" in a single glance.[41]

But there was yet another magnetic phenomenon that some believed was just as important: *inclination*. As Leibniz explained, magnets were also "inclined toward the horizon," and since these inclinations were also inconsistent when measured in a single location over time, it made sense to study this phenomenon too. Yet this required the use of a different kind of instrument, an "*Instrumentum inclinatorium*," Leibniz wrote, "so that *Inclinati Magnetis*, which is different from declination, could be observed."[42] He continued:

Inclination cannot be observed with a compass, but only with an inclinatory instrument, where the needle does not move in a horizontal plane, like with a compass, but rather in a vertical plane. This [instrument] shows that the needle stands neither parallel nor perpendicular to the horizon, but makes a large angle from the horizon, which is different in different locations.[43]

In a different context, Leibniz expressed his interest in the inclinator's mysterious tendency to rotate freely and spontaneously when allowed to do so; he saw it as way of accessing "a certain universal motion supposed to be in our globe."[44]

Gilbert, too, had described inclination as "magnetic dip" and advocated using inclinators or "dipping needles" to study the "remarkable motion of magnetick bodies dipping below the horizon by their own rotatory nature."[45] And by 1676, a British instrument maker and teacher of mathematics named Henry Bond had claimed that inclinators might be the perfect tools for measuring longitude. Two years before Bond made this claim, King Charles II had appointed a commission to test Bond's claims at sea, and several members of the Royal Society had become interested in testing inclinators. The fellows decided "to have a good inclinatory needle made and suspended in the Society's repository, to see what change there would be in it in tract of time."[46] Henry Oldenbourg, by then Royal Society secretary, even sent a copy of Bond's book to Leibniz.[47] But in the end, those involved grew frustrated with inclinators: they were harder to build than compasses and, overall, inclination proved harder to observe.

Declination, not inclination, became the geomagnetic effect that more people observed all over the world, because it seemed to offer an easier, more efficient way of explaining the earth's magnetic motions and secrets. When the Berlin Academy was founded in 1701, this became one of its early members' main priorities. Leibniz wrote that realizing the academy's mission and creating a network of interconnected academies was going to require the help of the tsar.[48] And he knew the tsar was very interested in geomagnetism:

It is well known that the Tsar, who loves the art of seafaring, sees the compass as the soul of the voyage; he knows that its needle always declines from true North and that this declination is different in different places. A yet to be solved enigma of nature is behind this, which, when solved, will mean a solution to the longitude problem and be a great help to pilots of ships.[49]

Several members of Berlin Academy responded to Leibniz's call to impress the tsar.[50]

When Leibniz finally met Peter I in person in 1711 (the same year Eberhard went to work for von Weyde), Leibniz proposed that the tsar sponsor a major research project that would involve studying the earth's magnetic field by gathering magnetic observations across the Russian Empire.[51] It was at this point that Leibniz mentioned the importance of studying both inclination and declination and the importance of equipping skilled *Observators* with inclinators. Leibniz told the tsar that he believed taking measurements with inclinators across Russia "would make it possible to learn magnetic secrets" and would give him a strategic advantage over others studying geomagnetism.[52] Leibniz knew he needed to act fast to win the tsar's support. He also knew that in Halle's Orphanage there was a small, steadily growing community of scientific observers in training who had precisely the kinds of expertise in mathematics and physics needed to undertake this kind of research project.

The introduction to Eberhard's *Attempt at a Magnetic Theory* notes that when he returned to Halle shortly after his meeting with the tsar (in Danzig) in 1716, he found his friend Semler had been thinking a lot about how best to observe magnetic effects too.[53] Semler may have discussed inclination and the *Instrumentum inclinatorium* with Leibniz in person during the latter's visits to Halle; like Leibniz, Semler was convinced inclination was an underobserved yet important phenomenon.[54] Semler showed Eberhard tables of observations gathered by Jesuit missionaries, and the two friends got to work; together they built an inclinator, and the following year, 1717, Eberhard left for Amsterdam, where he presented it to the tsar.[55] He then traveled to The Hague and demonstrated it before the Landgrave of Hesse-Cassel, a representative of the prince of Württemberg, a Dutch admiral named Baron von Wassenaer, and the Dutch natural philosopher Willem 's Gravesande.[56] He came away from these encounters so convinced he was onto something important that he headed across the Channel to London, where he published his new attempt at a theory, which included a "universal method" for measuring both longitude and latitude using a compass and an inclinator.[57]

Eberhard's method and instrument generated a great deal of interest in London, including offers of financial support from the king and efforts to conduct tests at sea.[58] According to Halle's chronicler, Johann Justus Dreyhaupt, members of the Board of Longitude were so impressed with Eberhard's ideas that they resolved to send twenty individuals who "understood the art of ships and mathematics" to investigate his claim. Isaac Newton appointed a deputy, William Whiston, to lead the investigation. Dreyhaupt noted that Whiston tried to talk Eberhard into turning over the discovery in exchange for sharing the prize with him. He was rumored to have met with Eberhard and to have written a report about it "as if he had discovered a new longitudinal method himself."[59]

Practicing Eclecticism, Theorizing about the Earth

The controversial mathematician and follower of Isaac Newton William Whiston did spend the most time with Eberhard during his time in London, and he recorded his impressions of their encounters.[60] In many ways, Whiston was an ideal contact for an Orphanage agent abroad: he was a supporter of Christian union, or irenicism; he had founded a society in London to promote the study of "primitive Christianity"; and he was a member of the SPCK. He was uniquely poised to appreciate the Halle Orphanage's status as a new kind of scientific community. However, he seems to have had no idea that Eberhard was affiliated with the Orphanage or the tsar at all. Remembering his first encounter with Eberhard, he wrote:

About the Middle of November, A.D. 1718, came to me one Mr. Eberhard, a German, born at Isleben in Saxony, the very town where Luther was himself born; who pretended to have a Method for discovering the Longitude by the Dipping Needle; which he had been several Months proposing to the other Mathematicians and Virtuoso's in Town; tho' without being able to satisfy any of them, that he had made any such Discovery.[61]

Whiston saw Eberhard as someone with no credentials and no institutional affiliation to speak of—entirely at odds with the culture of public science emerging in the city. This made him reluctant to trust Eberhard even though he was clearly quite interested in what he had to say. As he remembered:

I had no Notions of discovering the Longitude by the Way now propos'd. Nor was it possible I should have, since I did not then distinctly know what a Dipping-Needle was: nor indeed do I remember that, before That Time, I had ever seen such an Instrument in my Life.[62]

Intrigued, the first thing Whiston did was find a member of his own community whom he trusted and who could show him a "real" inclinator, since Eberhard's instrument, in his eyes, was useless. As he remembered:

I had a strong Inclination to see a real Dipping-Needle, and to try some Experiments therewith: for those that Mr. Eberhard had shewed me were so very small, that they were entirely unfit for Philosophical Experiments. Nor did I well understand, by what I saw there, even the common Nature or Use of such an Instrument. Accordingly, to satisfy my Curiosity, I soon enquired for such an Instrument and found that Colonel Windham had one, tho' it was small; which upon my Application he readily lent me; and as readily shew'd me the way of using it.[63]

Eberhard's decision to hide his ties to patrons, friends, and Halle in this new context meant that the instrument he carried was just as suspicious as he was.

But as we have seen, Eberhard *did* have credentials and an array of contacts and connections abroad; he had once served as an agent of the Orphanage in London and was in frequent contact with friends in Halle while living in Russia. Why would he suddenly pretend to be unaffiliated? One answer is that greed and hubris overcame him: Eberhard believed he had solved the longitude problem and wanted to keep all the glory and prize money for himself. Yet I propose an alternative explanation: by 1718 the Orphanage as a scientific community had changed too much for Eberhard's tastes, and he no longer wanted to be directly associated with it. The Orphanage had discontinued its efforts to actively pursue confessional reconciliation and promote the Orphanage in London—ties with the SPCK were mostly broken. Eberhard could not make his relationship with Peter I public knowledge because he really had no official relationship with the tsar. He knew the tsar through von Weyde, who had hired him because of his connections to Halle. Suddenly now Eberhard was alone with his inclinator in London.

Whiston was not only skeptical of Eberhard; he suggested he was an imposter who had stolen both the design for the instrument and all the observational data from a French Jesuit missionary named François Noël. Noël had spent fifteen years in the society's China mission, arriving in Macao in 1685. In 1710 he published a book containing "mathematical and physical observations" from China and India that was quite widely disseminated and speaks to his engagement with Kircher's geographical plan.[64]

In the introduction to his own 1721 treatise on the inclinator, Whiston explained that in 1706

Father Noël, of the Society of Jesus, when he sailed from Lisbon to the East Indies as a Missionary, had along with him a good Inclinatory Needle; and all the way made accurate Observations of the Angle of Inclination, as far as Fort St. George.[65]

Whiston remarked that Noël's "Set of Inclinatory Observations" were the best he had seen, "especially as they appear to have been carefully made, and with a good Instrument, and to contain the true Dip over so large a Part of the World."[66] He also described Noël as having "an Hypothesis of his own with respect to Two large Magnets he supposed to be within the Earth." He then asserted that Eberhard had simply adapted Noël's theory and was endeavoring to pass it off as his own:

Upon my Examination into [Eberhard's] Method, which at first I thought might admit of some Mystery, which he conceal'd; I found at last, when he sent me Pere Noël's Book, it was little more than that Author's Hypothesis, already mentioned, of Two Internal Loadstones and the Imitation thereof by the Insertion of small Loadstones under Maps and within Terrestrial Globes: With some hopes he had that Nature would afford a sort of Magnetick Needle, which should point East and West, as the ordinary ones pointed North and South. While yet he confessed he did not know that there was such a Power in Nature, as most certainly there is not: and yet without such a Power, all his Expectations must come to nothing.[67]

Whiston made a particularly bold allegation in this passage: not only was Eberhard pretending Noël's instrument and copious observations were his own, but he had also stolen Noël's theory—he seemed incapable of coming up with his own original ideas.

Not surprisingly, Eberhard's version of what happened between him and Whiston is quite different. He acknowledged that Noël's *Observationes* (published in Prague in 1710) inspired him to devise his own magnetic theory of the earth and admitted to relying on the Jesuit missionary's table of inclinatory observations, a standard practice.[68] He claimed that he had inserted small loadstones under maps and inside globes in order to demonstrate his magnetic theory, not as part of his method for finding longitude as Whiston later asserted.[69] Eberhard's theory actually grew out of his interest in Edmund Halley's hollow-earth theory. Yet Eberhard made it his own in order to use the copious inclinatory observations that he and Noël had collected during their respective journeys.

Whereas Halley had posited a hollow earth containing a rotating hollow globe with magnetic poles, Eberhard theorized that the earth contained two magnets, one located under the Americas and one located mainly under Europe and Asia. These "American" and "Eurasian" magnets were different sizes and rotated around two different magnetic cores at different speeds. Eberhard structured his theory around six axioms. He hypothesized:

1. That the Earth has very large caverns.
2. That in these caverns, among other things, are two large magnetic bodies.
3. That these magnets have poles, circumference and motion.
4. That these magnets can be distinguished from each other both by their size and their position.
5. That these magnets are the true and correct cause of all magnetic powers and effects.
6. That by understanding the very regular magnetic movements, the latitude and longitude of places can be found.[70]

Eberhard explained that he was interested in better understanding the cause of geomagnetic phenomena that no one (he mentions William Gilbert, Athanasius Kircher, and René Descartes) had managed to explain. Echoing sentiments voiced by Leibniz, he insisted that unless more observers attempted to make sense of existing magnetic observations, they would continue to hold no meaning.

Each of Eberhard's axioms was supported by a variety of proofs, which consisted of deductions made from observations and reasoning by analogy. For example, as proof of the first axiom, Eberhard pointed to ongoing discussions about the bodies of comets, which he said had a structure similar to that of the earth; he did not elaborate because, he said, he knew already that "the most famous philosophers of this island, namely Mr. Newton, Mr. Halley and Mr. Whiston, agree with me."[71] Considering all the different forms of creation living in the air, in the water, and on the earth, he reasoned, it was not a stretch to imagine different kinds of bodies living inside the earth's caverns.[72] Here is where we can begin to appreciate some of the more controversial components of Eberhard's theory, particularly his willingness to make use of analogical reasoning or logic that seemed to show a willingness to embrace the scholastic methods that had been rejected by his friends and mentors in Halle's theological faculty.

Eberhard very likely had encountered Christiaan Huygens's *Kosmotheoros* (1696) during his travels with von Weyde.[73] First published in Russian in 1717, the *Kosmotheoros* came on the scene as a kind of instruction manual: it offered an emerging educated public—interested in natural philosophical frameworks or syntheses—an introduction to "the various etiquette protocols used in Western European natural philosophy."[74] Its focus was on the universality of reason, including applying reason through analogy, which Huygens used to make a case for the presence of life on other planets.[75] The structure of Eberhard's theory is strikingly similar to the structure of Huygens's own, except that the subject matter is different: whereas Huygens focused on extraterrestrials, Eberhard focused on making a carefully reasoned case for large caverns, magnets, and life beneath the earth's surface.

To prove the other components of his hypothesis, Eberhard enlisted what he frequently referred to as "his guide, *Experienz.*" The real question, he wrote, was how many degrees lie between the poles of the two internal magnets and the poles of the earth. "After I asked my guide, *Experienz* for advice," he said, "I finally found a way to examine the boundaries of these poles and to recognize them."[76] He claimed to have determined that the North Pole of the European magnet was ten degrees away from true north and that the North Pole of the American magnet was only eight degrees away; additionally, the South Pole of the Asian magnet never went more than forty degrees away from true south and the southern pole of the American magnet never more than sixty degrees away.[77] "The truth of these things," he wrote, "is proved through the inclinatory needle [*Acus Inclinatoris*], that is, the upward and downward moving magnetic needle which, when oriented northwards, will never remain vertical or straight."[78] Eberhard included images of a compass and an inclinator in the description of his theory to help clarify the differences between the two instruments (fig. 17).

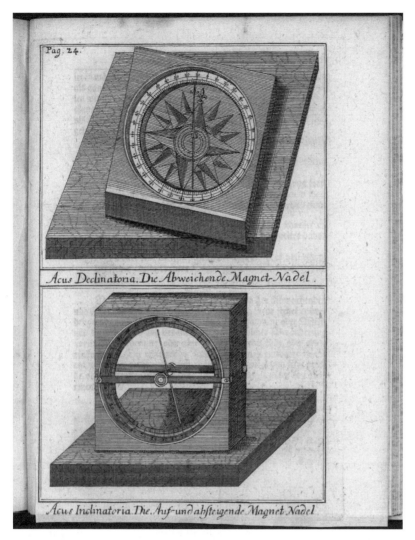

17 A compass and an inclinator compared, from Christoph Eberhard, *Versuch einer Magnetischen Theorie* [Leipzig: Martini, 1720], engraving. Credit: Niedersächsische Staats- und Universitätsbibliothek, Göttingen [8 ART ILL 2010 (3) RARA].

Another powerful way to understand Eberhard's attempt to create a new magnetic theory of the earth is to see it as an example of applied eclecticism in action. He crafted his theory by bringing as many perspectives, observations, and tests as possible to bear on the explanation he developed for the instability or unpredictability of magnetic phenomena. Yet his way of thinking about eclecticism corresponded more closely to

Wolff's aspirations for the *via eclectica*, not Buddeus's: his theory privileged reason over moral principles. It was overtly mechanistic, and it was ahistorical—no small matter in Halle by 1720. Even Whiston had found a way to make his own *New Theory of the Earth* compatible with a highly literalist reading of the book of Genesis. Eberhard, on the other hand, did not attend to origins. He did not situate his theory within a broader system of mosaic physics or consider the implications of his findings relative to contemporary cosmogonies or conjectural earth histories.[79]

As Claudine Cohen and Andre Wakefield explain in their introduction to Leibniz's cosmogony, a little text he called *Protogaea*, these earth histories were rooted in empirical observations and "included problems ranging from the origins of mountains and the causes of earthquakes to volcanic eruptions and the motion of the sea."[80] They were laden with "serious religious and philosophical implications," not the least of which was "the status of the biblical narrative as literal truth."[81] Yet another serious implication these conjectural histories raised concerned mechanistic frameworks for understanding the world, which were consistently being called into question by many of Halle's professors.

It is true that Leibniz's mechanical account of the earth's creation was more or less an extension of ideas Descartes had articulated in *Le Monde* (1633) and in a portion of his *Principia philosophiae* (1644).[82] But Leibniz's *Protogaea* also openly criticized the principles of Cartesian physics. In particular, it represented Leibniz's effort to address what he felt was Descartes's inability to reconcile his ideas with "a Biblical account of creation."[83] Cohen and Wakefield note that Descartes's decision to present his "history of the earth as a fiction or fable" also concerned Leibniz, "who considered such an attitude theologically dangerous and methodologically flawed"; he wanted to see a biblical account of creation used "as the framework for a real history."[84] At the same time, unlike Thomas Burnet's account in the *Telluris theoria sacra* (1680), Leibniz's account did not view specific events mentioned in the Bible as supernatural. Leibniz viewed the Great Flood, for example, not as a form of supernatural punishment or miracle but as an entirely mechanical phenomenon.[85] Eberhard's account was not only mechanistic and axiomatic; it evaded issues relating to the earth's origins in the interest of putting forward a new hypothesis about the underlying cause of geomagnetic effects on the earth's surface. This could not have impressed his friends in Halle, especially in light of the curricular revisions and growing concern about Wolff discussed in chapter 3.

Eberhard soon realized he needed to leave London: his theory and inclinator would not speak for themselves as he had hoped. He needed friends, patrons—a scientific community. Where could he go? As he was getting ready to leave, he received an invitation from Tsar Peter, who offered him twenty thousand ducats for the "famous instrument for measuring longitude" and invited him to come back to Russia.[86] But Friedrich IV of Denmark had also heard about the tests in London and managed to persuade Eberhard to come to Altona instead. Here he was given opportunities to improve the inclinator and to make more observations. It is not clear how long he stayed, but Eberhard eventually did return to Russia upon learning the tsar had drawn up detailed plans to send him as the leader of an expedition to the Kamchatka Peninsula. He intended to outfit him with a special ship, from which he would measure inclination along the entire western coast of North America.[87] After Peter I's sudden death in 1725, these plans were never realized.[88]

Sometime in 1719 or 1720, Eberhard returned to Halle, but he avoided the Orphanage. This time he went straight to Christian Wolff, who was serving as a member of both the Berlin and the Saint Petersburg Academies of Science. On June 15, 1720, Wolff wrote an evaluation of the inclinator for the Saint Petersburg Academy in which he explained that

because Pastor Eberhard requested me to circulate information about his discovery of a device for navigating at sea, I am willing to testify that, in my opinion, because of the variation of the vertical needle when pointed eastward, one can deploy the device:

1. if the needle is large enough so that one can accurately show the variation.
2. if through a straight line or through other observations it is known in advance how large the variation along each Meridian is and along each degree of latitude.
3. if it is proven why the variation in each place changes at a given moment of time.

And so, I suspect that with the instrument Mr. Eberhard has discovered investigations can be conducted at sea.[89]

Interestingly, Wolff stressed that the inclinator was only useful if the underlying cause of its inconsistent measurements in a single location was known, and he did not directly endorse Eberhard's efforts or his magnetic theory. He focused entirely on the potential of the instrument and noted it was Eberhard who "asked him to circulate information about his discovery." When he later published his *All Kinds of Useful Experiments* (1721–23), Wolff included a description of the inclinator in his chapter "On Magnets."[90] Also relying on François Noël's *Observationes*, he explained that magnetic inclination was a special kind of phenomenon that "has not been consistently observed since sailors were mostly trained to observe declination."[91] Nowhere in this volume did he allude to Eberhard, who may have become too controversial.

In the same year that Wolff was expelled from the University of Halle, 1723, the Orphanage press published Christoph Semler's description of the inclinator, including a three-part solution to the longitude problem. In this text, Semler mentioned Eberhard's involvement in the creation of this new instrument, but he suggested that his friend had not been able to demonstrate successfully how the inclinator could be used to measure longitude when he was in London.[92] According to Dreyhaupt's account, Semler deliberately "held back a few elements of the solution so that Eberhard would not be untrue to him."[93] Orphanage administrators and especially Semler tried to salvage their reputations by saying they knew from the start that Eberhard would take full credit for the discovery, so Semler deliberately did not share all the relevant information with him. As Semler said he had predicted, Eberhard "believed that he understood the thing entirely," and he "presented the solution to Parliament as his own discovery."[94]

By this point, Orphanage administrators were in an increasingly awkward position. Part of the Orphanage's ongoing appeal to outsiders was the access to scientific technologies and information that it offered to young people. Francke still wanted to attract the tsar and other powerful patrons to Halle, which meant he would have wanted to show that the Orphanage's *Observators* in training were studying pressing problems, like how to explain the cause of geomagnetism in the interest of solving the longitude problem. On July 7, 1716, around the time that Eberhard returned to Halle to collaborate with Semler, Francke reported in his diary that a prominent court counselor from Moscow had traveled to Halle "to see the Orphanage," was pleased with what he had seen, and planned to provide a full report to the tsar.[95] A few days later, on July 28, another associate of the tsar's, Count Alexander Golovkin, visited the Orphanage from Berlin; he agreed to carry several letters back to Russia for Francke and sent his son to study in the *Pädagogium* in 1717.[96]

At the same time, as the curriculum revisions from 1721 suggest, things were changing in Halle. Eberhard had crossed a line; administrators needed to find a way to make sure the inclinator remained associated with the Orphanage and not with him. They decided to publish their own description of the inclinator and method for finding longitude (fig. 18). Semler authored the text and noted that he had demonstrated the method in the Orphanage building while standing next to the Solomon's Temple model. This was his way of historicizing his efforts and effectively

18 Frontispiece showing an inclinator suspended over the city of Halle, from Christoph Semler, *Methodus Inveniendae Longitudinis Martimae* [Halle: Waysenhaus, 1723], engraving. Credit: Niedersächsische Staats- und Universitätsbibliothek, Göttingen [8 BIBL UFF 136/137].

locating them within a broader framework of mosaic physics and "pious" natural philosophy.[97] The method consisted of a three-part introduction to (1) using the vertical needle (*Acus Verticales*), (2) the unique dimensions of maritime environments, and (3) using solar clocks and other machines. Semler explained that the inclinator was one of many tools one might use to determine the underlying cause of geomagnetic effects, but he did not put forward his own magnetic theory.[98]

When King Friedrich Wilhelm I visited the Orphanage in 1720, he was told that Semler was the inventor of a new "device to measure longitude" that had been taken to England.[99] No mention was made of Eberhard at all. And it is at this point that Eberhard literally disappears from the historical record. By all accounts, he lived in Halle for the rest of his life (d. 1750), but I have not been able to learn much more about him. He had two sons, both of whom became scholars interested in magnetism and the mechanical arts. His youngest son, Johann Peter, held two professorships in Halle's medical and philosophical faculties, where he lectured on experimental physics (especially Newton's color theory, acoustics, fire-light and electricity, thunder and the northern lights) and applied mathematics (the art of building mills, hydro-technology, mining machines, and optics).[100] His oldest son, Johann Paul, became a private tutor and taught architecture and geography at the University of Göttingen.[101]

The inclinator (most likely a replica of the original) remained on display in the Orphanage museum, and a report in the 1730 edition of the *Halle Advertisements* indicates that Semler continued to demonstrate it publicly. By this time, he was allowed to offer his own mechanical collegium at the university using his collection of material models, machines, and instruments. These included "a new machine that can help determine the true cause of the ocean's tides," a material model of the city of Halle, a model of a salt mill, a model of a mine, and a singing clock. In the collegium, Semler also showed students a "magnetical experiment" involving a vertical magnetic compass (*magnetica verticalis*), which, the newspaper report indicated "still no one has attempted to use at sea for the good of the sailor and of which the same thing, when it is combined with the horizontal needle (*acu horizontali*) easily can be used to find longitude at sea using specially developed magnetic-hydrographical maps."[102] What were these special maps? They may have been maps showing "isoclinic" lines: a special kind of contour line that connects points where magnetic inclination is equal across the globe.[103] William Whiston produced one of the first isoclinic maps of southern England shortly after his initial meetings with Eberhard in London.[104]

Like many of those who had joined Catholic long-distance societies, particularly the Society of Jesus, Eberhard was from a modest socioeconomic background, flexible, talented, and willing to travel for Halle's Orphanage abroad. He was a skilled scientific observer who was able not only to construct and use specialized observational tools but to actually devise theories about the observations he had assembled. As an agent of the Orphanage in Russia, he was able to help Leibniz and Francke build relationships with the tsar and to explain the goals of the new scientific and educational community to those he met. He offered proof that the young men training and teaching in Halle were engaged in meaningful pan-European efforts to resolve unanswered questions about natural processes and pressing problems such as how to measure longitude. And they were able to do this because of the mixed training they received in practical mathematics (including how to gather observations with scientific instruments) and physics.

Yet even though Eberhard did what he was supposed to do—traveled widely, gathered and recorded observations, assimilated perspectives—as he was creating his own unique *Magnetic Theory of the Earth,* serious curricular revisions were under way in the Orphanage. These revisions sanctioned the rejection of certain perspectives along with the biblicohistorical literalism of an applied mosaic physics. As we have seen, this was a political move that was supposed to reinforce the authority of the theology faculty, including its professors' abilities to oversee who did and did not get to create explanatory frameworks, theories, or syntheses. Eberhard went too far when he created a theory that was axiomatic, overtly mechanical, and ahistorical. And in Halle's increasingly polarized environment, it now seemed as though he was working against his former friends and patrons in the Orphanage by insisting that reason, not moral principles, should serve as the foundation for an applied eclecticism. Orphanage administrators wanted to preserve their organization's reputation as a place of technoscientific innovation: indeed they wanted to be associated with the inclinator and its potential. But in the end, they chose to disassociate themselves from their star *Observator.*

SIX

Extending the Orphanage

Yes, I must tell my readers a little bit about how my faith in our institution has been powerfully strengthened. I have in my library not only Francke's news from the Halle Orphanage and *Pädagogium* but also Steinbart's news from the Zül-lichau Orphanage (and alone the foreword introduces still more of the same kinds of houses); I have every half year reports from the Orphanage in Cotbus; I receive news from the famous *Realschule* in Berlin and also reports from the Orphanage in Görlitz. Awhile ago a friend lent me the news reports from the new Orphanage in Grünstadt, founded in 1749. And even though I do not have the news reports from the Langendorf Orphanage near Weiâenfels, I still possess a short description of the life of the founder himself.   GOTTFRIED ZAHN, *ERSTE NACHRICHT*

The construction and further development of many Orphanages in different locations since the end of the last and the beginning of the current century has increased substantially. Through these many thousands of souls have come much earlier and much closer to the recognition of truth, which otherwise either would not have happened at all or would have been made to happen with great difficulty. Because one finds no traces of such healthy and useful institutions in the old and even the newest history books, reports are being written for us from all sides with news about them. . . . An Orphanage has been started up still in this century in *Augsburg* . . . . There is one in *Nordhausen*, headed by the preacher Herrn M. Kindervater, which now already has been standing in full bloom for a few years. And another foundation has been laid in *Wiese*, a village near the city of *Greiffenberg* . . . . One in *Langendorf* near *Weißenfels* has been started by a Christian and god-fearing farmer in reliance on the living God, whose plans have already been fully realized. One in *Königsberg*. Two in *Berlin*, of which the biggest was built at the cost of the king and the smaller at the cost of the widowed and recently deceased Frau Kommesserin. One in *Leipzig*. One in *Sorau*, which is taken care of by a dominion of counts. And the one that I should have named first, which was started in *Halle* by the honorable Prof. Francke in 1694, who was not carrying out his own intentions [but God's]. Because of its tremendous and varied uses [the Halle Orphanage] is known much more widely to the world and it is more beautiful than what I am able to express here. Finally, God also thought about our dear city *Züllichau* and prepared a clever instrument through many tests of faith, Sigismund Steinbarthen, . . . who was strongly awoken, in God's name, also to found [an orphanage], which in this historical news bulletin will be laid out, multifariously, before the eyes.   SIEGMUND STEINBART, *WARHAFFTIGE UND UMSTÄNDLICHE NACHRICHT*

Figure 19 offers two views of the Halle Orphanage facilities in 1749. The engraver, Gottfried August Gründler, depicted the main building as it appeared to observers as they approached the campus. The image below it provides an alternative view of the community's various components: the apothecary, bookshop and printing press, a variety of schools, the curiosity cabinet, and more.

In 1711 Francke described the Halle Orphanage as an archetype and remarked that there were already many attempts under way to start similar orphanages in other cities.[1] Leibniz also said he hoped this would happen in his correspondence with Francke. He aspired to create "a global network of scientific academies that would pursue coordinated research";[2] what emerged was a network of orphanages. Efforts to replicate the Halle Orphanage gained momentum mainly in the territories of Brandenburg-Prussia, yet there were also efforts under way in Hesse, Saxony, and Bavaria; parts of Russia; Denmark; and southeastern India. In North America, Renate Wilson has noted that "there were three specific and well documented attempts to replicate salient features of the Halle Orphanage . . . the Ebenezer orphanage in the Salzburger Georgia settlement, founded in 1733; George Whitfield's Bethesda Orphanage in Georgia founded in 1740; [and] the Economic Orphanage planned but never fully realized by Halle's emissaries in Pennsylvania . . . between 1750 and 1775."[3]

To bring my study of the Orphanage as scientific community to a close, I will focus on three types of replicas: orphanage-based institutions consisting of day schools (Langendorf, Sulechów, and Zittau), institutions that more closely resembled the *Pädagogium* (Königsberg's Collegium Fridericianum), and Johann Julius Hecker's "school of the real" (*Realschule*) in Berlin. Each of these replicas was unique, yet all of the founders and higher administrators maintained ties to the Halle Orphanage, generally modeling their curriculum after its schools. What follows are three short narrative vignettes that begin to explore these connections and that I hope will lay a foundation for further exploring the Orphanage's associated long-distance network.

Gift giving as a gesture of support and a mark of one's connection to the network was very important. Social anthropologists have long stressed

19 The Halle Orphanage facilit, 1749. Gottfried August Gründler, *Seitenprospekt oder Ansicht der Gesamtanlage des Waisenhauses*, engraving. Credit: Archiv und Bibliothek der Franckesche Stiftungen [BFSt/B Sb 0004].

how in a variety of cultural contexts gifts help generate and reinforce feelings of goodwill and communal solidarity.[4] Indeed, gift giving provided a way of building and maintaining alliances and reinforcing certain standards of conduct in the early modern republic of letters.[5] Yet it was also very often an exclusionary practice. Exchanging gifts helped forge meaningful connections between individuals, mostly men, with a certain understanding of themselves as scholars; it was a professional undertaking that involved an obligation to give and to receive. While many of those involved in the Halle Orphanage were also scholars, the organization aimed to extend the boundaries and activities of membership. Halle's Orphanage was a kind of meeting ground for individuals from across the social spectrum and, although there was an internal hierarchy ("higher" and "lower" schools, for example), there were new opportunities for collaboration while learning through recreation or exercise.

Generally the founders of replica orphanages invited all individuals, whether rich or poor, male or female, to support their institution by giving gifts. Donation lists from many of these institutions reveal a preference for giving books, models, and scientific instruments. Aspiring donors could also subscribe to newsletters detailing the progress of these institutions.[6] Giving a gift to a replica orphanage, or directly to the Halle Orphanage, was a way of participating in the growth of the entire network. Many apparently felt this philanthropic network was a cause worth supporting because each institution had something to offer. In many cases they provided young people with access to new information, technologies, and "useful sciences" in areas where very few other opportunities or schools existed. But perhaps even more important, many people reported spending hours observing these venues (and the teachers that worked there and even the children in some cases), much as one would venerate a shrine or relic, in the interest of his or her own improvement—and desire to become inspired.

The Orphanages of Langendorf, Zittau, and Sulechów

In the early eighteenth century, Langendorf, Zittau, and Sulechów were situated in borderland regions—places where political and cultural boundaries collided. The village of Langendorf is located just outside of Weißenfels, which was a small but vibrant seat of the Duchy of Saxony-Weißenfels, yet very close to Brandenburg-Prussia. Situated in a lush valley near the borders of modern Germany with Poland and the Czech

Republic, Zittau was also part of Saxony. And Sulechów was located in the Neumark, an eastern province of Brandenburg that now belongs to Poland. All three of these institutions were known as orphanages, even though children with parents from outside the city and region were allowed to attend school here.

Langendorf

In 1910, the director of the Langendorf Orphanage, Louis Bethmann, was invited to reflect upon the institution's two hundredth anniversary. He told the story of the farmer's son Christoph Buchen, who founded the institution on May 5, 1710. While employed as a servant in a Weißenfels guesthouse, Buchen formed a Pietist conventicle with three friends. Members of this group "took it upon themselves to collect some money and each time they came together, they laid aside a few coins—which otherwise might have been spent at the tavern—in a small box." Their friend Pastor Chryselius, "who preached the word in the spirit of Spener and Francke, often visited the boys and came to like Christoph Buchen especially, whom he encouraged to follow the wishes of his heart and to [take up the cause of] poor orphans."[7] Duke Johann Georg of Saxony-Weißenfels (r. 1697–1712) supported Buchen's plans and stressed his interest in ensuring that the institution would be selective, like the original community in Halle, and only take in "honest children" so that the orphanage could "realize its real purpose. These children should be allowed to come in," he wrote, and "be instructed in all the good sciences."[8]

Duke Johann Georg gave Buchen permission to build in a valley just outside the city and continued to find ways to support it. When his son Christian took over the duchy, from 1712 to 1736, his wife "donated amounts up to 200 *Talers* to the foundation often enough." Later, Bethmann noted, "at the cost of the duke, 11 orphans were raised in the foundation and 220 *Talers* were provided for them every year." He continued:

Hardly a day went by without some "Christian do-gooder" considering the Foundation in some way. To keep track of these daily offerings, Buchen kept a house book with an exact register. Also contributions poured in from far away, for example from Halle Missionaries in the East Indies, from Copenhagen, Vienna, Hamburg, Berlin, Augsburg, Gotha, Cöthen, Dessau, Weimar, Dresden, Jena, Merseburg, Naumburg, Weißenfels, and above all Leipzig. From Lisbon 294 Taler and 6 Groschen came in once and from London 100 Taler from an Earl's wife. Many gifts from the Halle Orphanage found their way to Langendorf and, as the house book demonstrates, were often delivered by Professor August Hermann Francke himself.[9]

Much of what is remarkable about the story of this orphanage is revealed in the passage above. Langendorf was (and is) not a cosmopolitan place; it is in the country, seemingly very far removed from corporations based in London, Amsterdam, or Copenhagen. Yet somehow this little institution managed to acquire donations from Halle missionaries in India, from Lisbon, and from an earl's wife in London. The Halle Orphanage sent gifts and, at least according to this report, Francke was a frequent visitor.

By 1714, several reports began to circulate describing the quality of the Langendorf Orphanage's daily pedagogical regimens and what forms of donations the house was receiving.[10] These reports were sent to benefactors on a subscription basis so that they could see for themselves how the institution was doing and what kinds of gifts it had taken in. The 1721 gift list included a six-volume set of Württemberg Bibles donated by a retired teacher, a small book acquired by a goldsmith from Naumburg during a trip to Switzerland, a Bible "with a preface written by Professor Francke" from a university student in Leipzig, and several musical instruments from university students in Jena.[11] The lists of received gifts are long and provide clear evidence of the willingness of even artisans and members of the lower orders to contribute to this orphanage.

Benefactor reports also described the Langendorf Orphanage as an "opportunity" for observers to learn and improve themselves by viewing the actual building. The 1721 report noted that the location and atmosphere of the institution was "beautiful"; it was positioned near woods and fields where "the children can be led to botanize in their free periods during the summertime." Some individuals had apparently complained that the orphanage should have been located in a city so that more people could visit it and "be inspired with a flaming love for the poor." But "because it lies at such a distance from the city, some say the work is robbed of its main purpose and that is, that it should be an object or opportunity to praise God."[12]

The children who attended school here were divided into groups according to their abilities and inclinations. Some were instructed in Latin; others learned Greek. Bethmann remembered that the quality of education boys could receive by the 1720s or so was at the level of what they could receive in a city gymnasium or even in the Fürstenschule Pforta, an elite *Adel* academy for boys from titled families not far from Weißenfels.[13] All learned geography, participated in singing periods every day, and on Wednesdays and Sundays "received instruction in arithmetic and musical instruments."[14]

A May 5, 1717, letter from Chryselius to Francke indicates that young people's progress through their daily exercises was carefully monitored with the hope that the most gifted could be sent to Halle. Chryselius wrote specifically regarding one boy, an H. M. Jüngling, whom he had at one point taken into the Langendorf Orphanage and found to be especially pious and promising, although he was a bit restless. If more efforts were made to help him develop an inner calmness, Chryselius wrote, he was confident that "something could still become of him" (*so könte doch noch was aus Ihn werden*).[15] Chryselius said he was pleased to learn that Francke had decided to accept Jüngling into the Halle Orphanage, which he described as a "temple."

Halle continued to give to Langendorf as well. In 1723, Johann Wilhelm Kruckenberg, a young theology student teacher from Halle, was sent to teach in Langendorf and would later direct the Langendorf Orphanage from 1730 to 1741.[16] Again, in this out-of-the-way place, Kruckenberg corresponded with Francke's son, Gotthilf August, about transmitting information and donations to missionaries in India. In a letter from 1739 Kruckenberg asked G. A. Francke to send him the latest missions report—and several copies at that.[17] Kruckenberg's post also included a portion of the donations he had received on site in Langendorf that he was sending to Halle so that Francke's son could decide how best to allocate the funds. In addition to Kruckenberg, the elder Francke sent Johann Michael Schumann to raise money for the organization by accepting more pupils whose parents could afford to pay for their education. In the "spirit of Halle," Schumann instituted collaborative meetings in which teachers conferred with the director of the educational foundation and with each other about "the art and manner through which the foundation could be improved."[18]

Zittau

The history of Zittau's orphanage is not as easy to trace as that of Langendorf; however, two individuals with ties to Halle were active there in the institution's early years. Gottfried Hoffman corresponded with Francke from his post in Lauben before moving to Zittau in 1708, and Martin Grünwald (who had also attended gymnasium in Zittau) encountered Francke in Leipzig, where he first tried to start an orphanage in the nearby city of Bautzen and then moved on to Zittau. Grünwald's descriptions of the Zittau Orphanage made use of the same rhetorical strategies we saw operating in Langendorf. They constantly hold up the aesthetic qualities of the model building, including its ability to house living model children. Grünwald noted it was built on the site of a cloister that had at one time housed monks of the Celestine order. It was "almost in the middle of the city" next to the main church. Its proportions were "perfect": the building was structured according to the "newest architecture" and was "not only useful and considering its purpose, very comfortable, but also constructed very gracefully."[19]

Zittau orphans' practices of piety begin with a brief introduction to their early morning prayers, like those that the monks who inhabited the space before them had engaged in. They began their exercises by singing a "morning song," asking the blessings of Martin Luther on their house, reciting the main piece of a catechism, offering "real sighs" imploring the divine to "accomplish something inside" of their bodies and to open their eyes, praying, reciting their "rules for living," and singing more songs.[20] At this point other children, whose parents wanted them to be educated at the orphanage but who lived at home, were allowed to enter. All children were constantly reminded of the institution's main purpose, which Grünwald described as "instruction in true and living Christianity." To realize this purpose, they received training in practical mathematics.[21]

Grünwald's description notes that several Zittau orphans went on to become apothecaries, sculptors, bookbinders, printers, instrument makers, and goldsmiths and that the orphanage's training regimens prepared them for entrance into these careers. The vocational training one received was "oriented around the talent of each child," he explained. And since the children were allowed to occupy themselves however they wanted in their free periods, they would be more inclined to discover on their own what they were good at doing. "If during their free periods and of their own accord through making horizontal projections painting, cutting and building they discover their own inclination and inborn dexterity," Grünwald wrote,

then they are given the opportunity to learn mathematical principles, including how to use the Circul, the Lineal etc. Special periods are set aside wherein they should also act out all of the various and curious examples from mechanics so that, after some time, they learn the names of the instruments, including the technical terms and special turns of phrase that artists and craftsmen need and use. It can only be God who has made the usually selfish and superficial members of wealthy society willing to prove their good will by inspiring certain people who are experienced in the free arts and the handiwork of mechanical exercises to make accurate models and other helpful aids available to be grasped by the poor. It should not be doubted that in our orphanage, poor children carve wood and prepare tools whose varied uses will serve many purposes in the future world.[22]

Donors performed philanthropy by giving models and "other helpful aids" to children of the lower social orders. Young people training here were invited to acquire expertise and instruments that could serve some specific purpose.

Sulechów

A leading center of Brandenburg-Prussia's cloth-making industry, the (now) Polish city of Sulechów became famous for the orphanage founded there "in the spirit of Halle" in 1719 (fig. 20).[23] Like Christoph Buchen, the "poor farmer" who had been inspired to build the Langendorf Orphanage, a poor tailor named Siegmund Steinbart is still remembered as the founder. By the end of the eighteenth century, his facility included an orphanage, day and boarding school complexes, a *Pädagogium*, and its very own royal academy for training teachers. Steinbart's son Johann Christian attended school in Halle's Orphanage, studied at the University of Halle, and later returned to Sulechów to direct the model community there.

Although eulogized as poor, the elder Steinbart did quite well in his profession. He held a prominent position in a cloth-making firm that had donated money to the Halle Orphanage and became the point of contact for this transaction.[24] In a gesture of gratitude for the generous donation, Francke invited Steinbart and his wife to visit Halle in 1716.[25] Steinbart could not make the trip then but sent his son to be educated in the community's Latin school the following year instead. A few years later, in 1719, he was able to travel to the Leipzig book fair and stopped in Halle to see the Orphanage. Legend has it that he was so inspired by his experience there that he decided to start an orphanage upon returning home.

Sulechów's orphanage was declared a "public work" that had been funded entirely by God, even though King Friedrich Wilhelm I gave it money and helped to supply the building materials.[26] Steinbart obtained permission for his institution to acquire its own printing press and bookshop and began receiving large donations from prominent Prussian families—including the wife of General Friedrich von Derfflingen, who sent 1,000 taler in 1728. This same benefactress would later donate an additional 6,000 taler to the institution and set up a scholarship fund for theology students.[27] Elite Prussian officers who preferred to send their boys there loved the institution and lavished it with all manner of gifts.

Like most of the institutions founded after Halle's example, yearly reports listed the various donations that individuals were sending from near and far. In 1730, Steinbart noted that the king had given the institution a mine and that benefactors from Halle had sent an array of

20 Sulechów's Orphanage; frontispiece and title page from Sigmund Steinbart, *Wahrhafftige und umständliche Nachricht* [Berlin, 1723], engraving. Credit: Archiv und Bibliothek der Franckesche Stiftungen [BFSt/B 129 E 14].

medicines.[28] Commenting on the receipt of one hundred taler from three preachers, Steinbart exclaimed: "Father . . . your love for the poor not only strengthens what you are attempting with this orphanage but also in other institutions, especially in Halle and . . . the Evangelical Danish mission in Tranquebar."[29] A 1731 report also explained that the institution was attempting to develop its own collection of objects. "Regarding a collection of *naturalia*," Steinbart explained, "so far we have had a slow beginning":

Three years ago I brought back from my trip to Austria an artful . . . chain, like the ones that very pious Catholics place upon the naked body. . . . A little over a half year ago, a friend gave us a model of a mine contained in a sealed glass bottle, in which all the different mine-related occupations are represented. In 1730, an honorable philanthropist donated a Turkish spoon. In addition to this . . . several different benefactors have offered up different kinds of curious *naturalia* that they maybe have more than one of. . . . We would be very grateful towards those who might be willing to send similar kinds of things to us.[30]

Berlin's "School of the Real"

Johann Julius Hecker came to Halle in May of 1726 to study theology and ended up working for six years as a teacher in the *Pädagogium*. His biographer notes that he learned all of the "most important forms of knowledge" from his teachers Justus Breithaupt, Paul Anton, August Hermann Francke, Joachim Lange, and Johann Jakob Rambach of the theology faculty of the university.[31] But while teaching in the *Pädagogium*, Hecker befriended professor of medicine Friedrich Hoffmann and became increasingly interested in the study of medicine, anatomy, and botany. Hecker taught languages but wrote a textbook on anatomy and another on physiology and botany, both of which were published by the Halle Orphanage in the early 1730s.[32]

In 1735, after a short tour of the Netherlands, Hecker worked in Potsdam's "military orphanage" as an inspector and pastor. In 1739, he moved to Berlin to preside as the pastor of the new Church of the Trinity (*Dreyfaltigkeitskirche*), which had been founded by the king the year before in a section of town becoming known for its small manufactories. It was here that he received permission to start a mathematical "school of the real" (*Realschule*), or universal school, where soldiers, artists, craftsmen, and servants were to be educated alongside children from elite and middling families (fig. 21).[33] "A school of the real is a school for everyone," he

21 The building of the Heckersche Schulanstalten in Berlin, engraving. Credit:
  Kupferstichkabinett, Staatliche Museen, Berlin [Kart. Y 46555], courtesy of Art Resource, NY

explained, "the intellectual, the artist and the craftsman, who all must be raised according to their future purpose."[34]

Hecker extended Semler's efforts to make objects the focal points for lessons in any subject in his *Realschule*. With the help of his friend Johann Friedrich Hähn, who had joined him as an inspector of the new complex in 1753, he built a model and machine room consisting of "the most exact models of machines, buildings, columns and famous paintings from antiquity."[35] "Some people think this thing is a big joke [*Spielwerk*]," Hecker remarked, but he persisted in his efforts to transform the school into a place where young people received useful information and exposure to new technologies.[36] He divided it into three components: a "boarding school organized like a *Pädagogium*," a Latin school ("nothing other than a language and arts school"), and a German school ("actually a craftsman's school").[37] There were a total of eight theology classes in the complex, six Latin classes, two Greek and Hebrew, six geography and history, seven calligraphy, six orthography, three epistolography, two mathematics, seven arithmetic, four drawing, and two "orders" for the practice of vocal music. Individual lessons were offered as needed in botany, anatomy, the science of manufacturing and small proprietorship (*Handlungs und*

Manufacturwissenschaft), mining (*Bergwerkskunde*), mechanics, and the "science of natural and artificial objects." More than thirty teachers reportedly "busied themselves with well over 1000 pupils" in the earliest years of the school.[38]

By the 1740s and 1750s Berlin's *Realschule* had in some ways surpassed Halle's Orphanage in terms of its ability to offer young people a different experience than they could have in other kinds of schools. The philosopher Christoph Friedrich Nicolai studied in the Halle Orphanage before moving to Berlin to attend the *Realschule*, and he remarked that the focus in Halle had reverted to learning "Latin and Greek words" along with "an entirely incomprehensible and backward Geography."[39] He also lamented there was a striking lack of attention to "mathematical or physical concepts of the Earth" there;[40] this made his transition into the *Realschule* especially memorable. As he explained:

I arrived in the "school of the real" in an entirely new world. Everything I had learned in my previous schools was uninteresting and not meaningful; on the contrary, everything I learned here seemed interesting and varied so that I could not contain my excitement in the first few months. . . . In the study of nature some kind of universal idea would be suggested and then many tests would be carried out. There was an air pump . . . a barometer and thermometer, as well as other physical instruments. Electricity was then something really new and we were actually taken to a place where there was an electrical machine and we saw all the latest experiments. Especially interesting was the manufacturing class. All the masterpieces of the craftsmen in Berlin who wanted to become masters were shown to us and, twice a week, the entire class was taken around to all of Berlin's manufacturers and factories, always following a certain, prescribed order. Each one of us had to describe, in our own words, what we had seen.[41]

Hecker's school became a place to conduct research. It was also a meeting ground for individuals from across the social spectrum who learned through object-based pedagogies. Even university-bound scholars used their hands and collaborated with the sons of craftsmen, artists, and engineers.

Königsberg's Collegium Fridericianum

A preacher's son named Theodor Gehr founded Königsberg's Collegium Fridericianum in 1698. Originally from Königsberg, Gehr studied theology and law and worked as a teacher before he met Spener in Berlin in

1689 and became involved with Pietist conventicles. From 1694 to 1697, two theology students from the University of Halle spent time in Gehr's household as private tutors. Inspired by the examples of these young men, Gehr took his entire family to visit the Halle Orphanage in 1697, where he collaborated with Francke and another colleague in the theology faculty, Johann Justus Breithaupt, for approximately one year. Upon returning home, he continued to correspond with Spener and Francke and became consumed with the idea of creating an institution in Königsberg modeled after Halle's.[42] When he surveyed existing schools in his city, he concluded that "lessons were entirely grammatical and rhetorical and offered preparation mainly for academic disciplines. . . . Real knowledge was hardly taught to pupils."[43] Also, a strange pattern had developed whereby many teachers were offering basic lessons in their "open" periods but saving some information for young people whose parents were willing to pay them for it. Gehr complained that the quality of the education one could receive in these schools was so bad that several Prussian Protestant families had begun sending their sons to Jesuit schools in the Ermland (a region of eastern Brandenburg-Prussia). It had become so common, in fact, that the Prussian king declared it illegal.[44]

Francke and Gehr maintained and expanded their relationship in part by sending gifts to each other. In February of 1698, Gehr sent Francke a gift for the Orphanage's curiosity cabinet.[45] In June of 1698, Francke sent a Halle University student named Georg Christian Adler to help Gehr with his own efforts. Adler arrived, carrying with him a copy of one of Francke's first descriptions of Halle's Orphanage, which he shared with all of Gehr's friends and acquaintances. A few days after Adler's arrival, Gehr wrote to Francke that the two had started a vernacular school for young girls (all lessons were in Polish) and were struggling with getting the private Latin school off the ground since founding a new school generally was a privilege granted only to city authorities; Gehr also conveyed his plans to build a *Pädagogium* modeled after Halle's.[46] The two finally managed to open a small Latin school that provided lessons in Latin, geography, history, and science (*Wissenschaft*) "because it is not too difficult for the children but rather is fun and pleases the spirit."[47] They continued to struggle against civic authorities who said the school was illegal because the two had not acquired permission to build it through the proper channels, but in 1701 a royal privilege was finally granted, allowing Gehr and Adler to continue their work.[48]

Gehr traveled to Berlin and Halle again in March 1701 to ask for more advice about how to go about expanding the ensemble. Several second-hand reports suggest that at this point friends and allies in Berlin seemed to have lost confidence in Gehr's ability to direct the fledgling institution. Spener threw his support behind Heinrich Lysius, a scholar who had spent time in Halle with Francke and Spener as early as 1694 when efforts to build Halle's Orphanage were first getting off the ground. Lysius had expertise in a variety of areas, including ancient Greek, Latin, Hebrew, and Aramaic, as well as mathematics, geometry, astronomy, philosophy, and theology. After returning from a tour of Denmark, Sweden, and Norway in 1701, he learned from Spener of the need for more help in Königsberg. He went back to Halle, acquired a doctorate in theology, and then became immediately involved with efforts in Königsberg as the appointed director of its *Pädagogium*.[49]

After Gehr's death in 1703, Lysius took over the Königsberg organization. He remained in close contact with Francke, who sent him a former student from his "oriental seminar," Abraham Wolf, to help out where needed.[50] Wolf had just returned from a mission to Astrakhan, a multi-ethnic city of merchants near the Caspian Sea that was then part of the Russian Empire. He had been forced to return early because of an outbreak of plague there, only to arrive in Königsberg to find the city struggling with the same problem. Unlike most of those teaching in the city, Wolf was willing to stay on. To this day he is remembered as a kind of hero, who kept the struggling school afloat when it was on the verge of failing. Francke's and the king's ability to help Lysius pay off the organization's steadily accumulating debts probably helped even more, however. In 1718, Lysius received funds to improve the school's buildings, which included repairing the roof, adding a new floor with more rooms to house pupils, and building an observatory on top of the building, which was outfitted with new telescopes.[51]

In 1729, Georg Friedrich Rogall, another friend and former teacher from the Halle Orphanage, took over the Königsberg organization. Rogall had originally come to Halle in 1722 to study with Christian Wolff, and during his tenure in Halle he befriended most of the members of the theology faculty as well. He became increasingly involved with Francke and the Orphanage and working as a student teacher there in 1723. At this point, the Königsberg complex had increased its visibility as a collegium, but back in Halle, there was concern that the organization had lost sight of its original mandate. Rogall was supposed to use all his energy to "keep these institutions in step with the Halle Orphanage," which he did by growing the facility's vernacular schools and starting a program like Halle's for employing more students from the university as teachers in exchange for room and board.[52] Rogall corresponded frequently with both Francke and his son, Gotthilf, throughout his time as director. Missionaries in India—namely Benjamin Schultze, Nikolaus Dal, Christian Friedrich Pressier, and Christoph Theodosius Walther—were also kept informed about Rogall's work in Königsberg.[53] When Immanuel Kant enrolled in 1732, Rogall was still the director.

Rogall's replacement was a former Halle University student named Franz Albert Schultz, who, like Rogall, had studied theology but also frequently attended Christian Wolff's lectures in mathematics and philosophy.[54] Schultz later wrote Gotthilf Francke to say he had become deeply engaged with Wolff's philosophy in Halle, to the point that he felt that he too might have been kicked out of the city if people had known about it. He was given permission to offer lectures in mathematics and philosophy and remained loyal to Wolff, even when he was asked by Francke to use his influence in Berlin (at this point he was working in a military school and had friends with direct connections to the court) to further diminish Wolff's reputation.[55]

Under Schultz's leadership, the Königsberg community produced several fairly detailed descriptions of its curriculum and moved again toward investing in its *Pädagogium* in particular. In his *News about the Current Institutions of the Collegium Fridericianum* (1741), adjunct director Christian Schiffert explained that the king had requested a reorientation of the Königsberg facility. He wrote:

Now it is important to note that according to the best intentions of his majesty, these institutions have been arranged, as much as possible, to follow the example of the *Pädagogium Regium* in Halle; one has also written this report according to the same style of the description of the latter . . . so that the similarities between our facility and the Halle facility will appear even more clearly before the eyes.[56]

Although the course offerings were not as varied as in Halle, there was a noticeably higher emphasis placed on geography, history, and practical mathematics—in keeping with the changes described by Freyer in the 1721 curriculum revision.[57] Schiffert wrote a special geography textbook. It included a short introduction to mathematical geography and the geography of the globe, including practical advice for measuring long and short distances and determining the time in particular locations. Advanced pupils were shown a variety of maps and even trained in the art of mapmaking. Additionally, the collegium offered two mathematics classes. In one class,

the easiest truths of mathematics that can be experienced through mechanical evidence and tests are demonstrated; in the second class, pupils are taught using the first part of Christian Wolff's *Beginner's Guide to the Mathematical Sciences*, namely arithmetic, geometry and trigonometry. . . . Through their studies they are led to become competent and adept at other sciences as well.[58]

Schiffert explained that the teachers were expected to make their lessons "as lifelike" as possible and to teach in a conversational question-and-answer style. Whenever possible they were to teach in the presence of objects, prompting their students to ask "What is that?" and "How does it work?" and to find out the answers for themselves.[59] Gifted pupils were encouraged to attend philosophy collegia at the university if they were so inclined; however, unlike in Halle in the 1720s and 1730s, where Buddeus's textbooks on *Eclectic Philosophy* were required reading for those chosen to study advanced philosophy, at Königsberg Wolff's writings "on metaphysics, natural theology, moral philosophy and physics" were the main texts teachers and their pupils consulted.[60]

––––––––

In some ways, exactly what Buddeus and other theologians in Halle were worried might happen actually did occur in the 1720s and 1730s, in an institution that was entirely modeled after their own. The tenets of Wolff's philosophy were being taught at a replica institution in Königsberg. Wolff's confidence and charisma also appealed to Halle's university students and to young people more generally, who traveled from far and wide to hear his lectures. Voltaire made note of Halle theologians' jealousy of Wolff's popularity with students as a deciding factor in their decision to have him removed from the city. "Wolff, I must tell you, attracted to Halle a thousand students from every nation. There was in the same university a professor of theology named Lange, who attracted nobody; in despair at freezing to death alone in his lecture hall, he quite reasonably decided to ruin the professor of mathematics; following the custom of his kind, he promptly accused him of not believing in God."[61]

The Pietist theologian Joachim Lange, mentioned above by Voltaire, did play a major role in maneuvering to have Wolff expelled from the city.[62] It just so happened that he also had a son, Johann Joachim Lange, who directly benefited from his father's interventions: the moment Wolff left the city Lange's son Johann took over the suddenly vacant mathematics professorship. But he was hardly qualified for the position, having produced only a dissertation on the translation of the Old Testament from Hebrew into Aramaic, a six-hundred-page book on New Testament grammar, a dissertation about the origins of doctors in Egypt during the era of the first

Christians, and another dissertation on a Jewish sect known as the Essenes.[63] Yet Lange had a special kind of expertise in biblical history and philology; he had trained at the Orphanage and had the support of the entire theology faculty. He was not interested in upending existing disciplinary hierarchies at the university, believed in combining scholarship with practical training in the mathematical arts, and was able to offer collegia in experimental physics at the university—as Wolff had done as well.[64] In 1735, the Orphanage published Lange's textbook on dogmatic and experimental natural philosophy and, several years later, used his introduction to the basics of Linnaean systematics to reorganize the Orphanage's museum.[65]

Despite all that had happened, including the turn toward mosaic physics and marked turn away from someone with a different point of view, the Orphanage continued to function as a "scientific community" inhabited by student teachers, pupils, and professors who were deeply engaged with the most pressing philosophical, social, and political issues of their time. In 1719, Francke sent some advice for those involved in the "Project to erect an Orphanage" in Tobolsk. He recommended that the young people there be taught geography and history, especially "the history of the beginning of the world from the Bible," but he hoped they would also learn geometry, which he described as "the first of all sciences."[66] He continued to work with Christoph Semler on a variety of projects, including starting a manufactory on the grounds of the Orphanage for producing globes and other models of the solar system.[67] The organization continued to send *Observators* into parts of Russia, and Halle-trained missionaries in India even got involved in ongoing species designation projects.[68] Some of Germany's most famous intellectuals, including Alexander Gottlieb Baumgarten, trained in the Orphanage; Baumgarten relied on Francke's understanding of "aesthesis" to help him create a new theory of moral education and aesthetics.[69]

Writing in 1754, Gottfried Zahn, the director of an orphanage modeled after Halle's in Bolesławiec (Bunzlau), said he was surprised how many people thought that "after the Halle Orphanage no similar institution has managed to continue, but rather their good beginnings are now disregarded and they have all gradually slipped away."[70] Indeed, he said that some even argued the "orphanage period" (his phrase) was over and that "all the amazing things these institutions achieved have ended."[71] Zahn discounted those who said these things by claiming that simply observing any one of these institutions offered "indisputable proof" of their lasting impact. But he understood that the Halle Orphanage's reputation had been damaged.

In the early years, when Francke began to collaborate with Leibniz and Tschirnhaus after having been forced out of the city of Leipzig, he was a rising star, and the Orphanage was a cutting-edge institution. Finally the school had become a place of scientific research. By then there was a major state-supported institution in Brandenburg-Prussia that was training skilled observers with expensive, highly coveted forms of instrumentation—burning mirrors, air pumps, barometers, inclinators, and more. The organization was sending out these skilled observers to study natural processes, solve pressing problems, and gather information abroad. After the Wolff controversy, though, the Orphanage and the Pietist theologians associated with it were increasingly ridiculed; the links between Pietism and dangerous, irrational enthusiasm that Francke had tried to overcome early on in his career returned with a vengeance. Wolff fanned the flames by referring often to "the Pietists" as a dangerous collective and describing how they had conspired against him in Halle. He continued to raise questions about theology's status as the highest faculty in the university, and in the end he won. In 1740, he was invited back to Halle by Friedrich the Great and stayed there for the rest of his life.

As Jonathan Sheehan has explained, "Already by 1736, a certain cultural picture of the Pietist was ripe for caricature by writers like Luise Gottsched, whose *Pietists in Petticoats* [*Die Pietistery im Fischbein-Rocke*] mocked an overly enthusiastic mother." He continues:

Having fallen into a religious faint, she needs only her daughter shouting in her ear "Arnold! Petersen! Lange! Gichtel! Francke! Tauler! Grace! Rebirth! The inner spark!" to rouse her. *And for the most part, modern scholars have relied on this cultural picture as well.*[72]

Sheehan notes that despite scholars' efforts to stress the "more objective developments within Pietism—the economic theory that structured the Halle community, the concrete organization of time for students," they continue to embrace a particular way of thinking about an enthusiastic "Pietist ethos" or to identify peculiarly "Pietist" forms of cultural production. This has led to a tendency to accept, often without question, a generic way of understanding Pietists as antirational and anti-Enlightenment, as "individualistic, inward looking, at times otherworldly, disinclined to political activism."[73]

Like Sheehan I have tried to respond to these issues by focusing on what those involved with Halle's Orphanage actually did and on the lingering tensions and ambiguity surrounding the very use of the word *Pietist*. My findings point to the real ways in which Halle's Orphanage filled a much larger need in central Europe for semipublic venues to hone and refine experimental and observational procedures by acquiring expertise in practical mathematics and physics as part of a collective. This does not mean that Halle's Pietist theologians were not interested in individual religiosity, in the cultivation of *pia desideria* through spiritual exercise and the emulation of virtuous archetypes—they were. But their end goal was to generate higher forms of affective or emotional intelligence, not enthusiasm. They drew from a variety of perspectives presented in manifold discussions at the university (and beyond) about how exactly the affections and cognition collaborate as one observes, apprehends, and synthesizes information.

This last point is worth emphasizing, particularly given how closed-minded and conflict-ridden Halle's university, and particularly the theology faculty, were often described and later remembered. The University of Halle and its "universal seminar," the Orphanage, were first founded as spaces that might help generate new modes of sociability predicated on a commitment to taking seriously all perspectives on a particular issue or problem. Underlying the organization's status as a space for promoting collaborative research was a sustained effort to institutionalize eclecticism as a scientific methodology. Proponents of eclecticism, such as Johann Daniel Herrnschmidt, believed in empowering certain individuals—particularly those who had been judged able or gifted by their teachers—to conduct their own investigations of nature and to make original contributions to ongoing debates and conversations.

A key part of educating individuals capable of doing this meant teaching them "how to see," helping them become skilled observers by presenting them with a variety of models placed "in perspective," and providing them with the specialized equipment they needed to see *into* things, to see *all at once*—or even to observe invisible phenomena. I have stressed the fundamental and lasting impact of Tschirnhaus's ideas on the Orphanage schools' curricula in this regard, particularly his emphasis on the need to reconcile empiricist and rationalist approaches to the study of nature and to attend to the world's "three essences": *rationales, reales,* and *imaginables*. Orphanage teachers and administrators were especially interested in the potential of material models, especially the organization's enormous model of Solomon's Temple, to inspire a form of "conciliatory seeing" in young people that was simultaneously rational, affective, and anchored in a familiarity with real things in the world. The temple model was beautiful and impressive; it inspired and triggered the kinds of pious desires Pietist theologians were so interested in. But it was also rationally constructed, requiring expertise in mathematics (especially geometry) and an effort to learn as much as possible about the real or original Temple.

One accessed the real Temple by opening oneself up to and internalizing myriad perspectives, a crucial part of embracing the *via eclectica*.

These efforts to standardize eclecticism as an actual method or way of coming to know the world were ideally suited to helping further the goals of those associated with Brandenburg-Prussia's irenical turn. Yet as we have seen, efforts to move beyond tolerance and to unify confessions ultimately failed and contributed to a reconfiguration of the ideals that led to the construction of the Orphanage. The process of reconfiguring these ideals involved a substantial curriculum revision and a reevaluation of the very meaning of the term *eclecticism*. Halle's theologians increasingly clung to the way of thinking about eclecticism promoted by Buddeus, who insisted that moral principles, not choices structured by logic or reason, ought to serve as the foundation of syntheses and theoretical frameworks. On these grounds, he insisted, it was permissible to reject certain perspectives entirely. As a result of this shift in focus, the community began to reject some individuals who had once helped to build the Orphanage's reputation abroad. Orphanage administrators went out of their way to remain associated with the inclinator, for example, while at the same time disassociating themselves from the *Observator* Christoph Eberhard, who arguably knew the most about it and had once been closely associated with the community.

Although Orphanage administrators cast aside their original mandate to train eclectic observers who moved beyond "mere toleration" and found the good in different points of view, their organization continued to serve as a model community long after the period under consideration here. Indeed, there was a fundamental correspondence between the quality of "moral-theological" education one could receive in Halle by the middle of the eighteenth century and the emerging theories of "philanthropic happiness," or *Menschenfreude*, that later inspired Johann Bernhard Basedow to start a new "school of philanthropy," or *Philanthropinum*, in Dessau (founded 1774). Hanno Schmidt finds evidence of this link in the notes of Joachim Heinrich Campe, who, while listening to a theology lecture in Halle in 1768, wrote that "in the drive for happiness, which we all experience, the most important question that a rational person must ask of himself is: 'how can I be happy?'"[74] "The study of theology in Halle," Schmidt concludes, "quite certainly led a larger part of the theology students into the intellectual neighborhood of philanthropy."[75]

Basedow was not interested in *philotheïa* or *philanthropia dei*, two terms Francke had used in his early writings on love; instead, he said that in his school young people cultivated *philalethie*, or the love of truth. He wrote

22 Daniel Chodowiecki, *Unterricht im Naturalien Kabinett*-Schüler und Lehrer (lessons in a naturalia cabinet with pupils and teacher) from Johann Basedow, *Elementarwerk*, plate 48 [1774], engraving. Credit: Kupferstichkabinett, Staatliche Museen, Berlin, courtesy of Art Resource, NY

about this love in a treatise he called *Philalethie: New Perspectives on the Truths and Religion of Reason*.[76] In it, he referred to himself as a "friend of truth" and described his strategies for seeking it. Figure 22, from Basedow's *Elementarwerk*, offers perhaps the most powerful example of how he endeavored to generate *philalethie* in his pupils. In the image, a teacher stands in the middle of a room surrounded by drawings of *naturalia* and mathematical instruments. He points at a sign on the wall that reads "Book of Nature and Customs, Book of Religion." As Anke te Heesen has explained, this particular image accompanied a lesson in the *Elementarwerk* about the "virtue and blissful happiness of children."[77] When children loved what they were experiencing and were allowed to experience it "in a natural sequence," the hope was that they would understand the relationships between these things and learning would be *ludus*. In his school, Basedow continued to explore links between affect, observation, inspiration, and investigation in a way that would have seemed very familiar to those involved in the original Orphanage—even when other institutions, such as Hecker's *Realschule* or the *Philanthropinum*, had become more popular.

Immanuel Kant, who attended school in a *Pädagogium* replica institution (*the* Collegium Fridericianum), raised funds for the *Philanthropinum*. He encouraged all "noble-minded people in all countries" to support the "nearly perfect" school by subscribing to the newsletter *Pedagogical Conversations*:

Subscriptions to the monthly publication of the Institute of Dessau, entitled *Pedagogical Conversations*, are now being received at the rate of 2 Reichstaler 10 Groschen in our money. . . . It would perhaps be best (but this is left to individual discretion) to devote a ducat, in the way of subscription, to the furtherance of this work, whereupon the surplus would be refunded to everyone who demanded it. For the institute in question flatters itself that there are many noble-minded people in all countries who would be glad of such an opportunity to add, at this suggestion, a small voluntary present to the amount of their subscription, as a contribution to the support of the institute which is nearly perfect, but which is not being helped as much as had been expected.[78]

Despite the clear resonance between the Orphanage and the *Philanthropinum* as institutional archetypes, by the end of the eighteenth century Dessau's *Philanthropinum* had become the new "model school." There were no theologians involved; in fact Basedow renounced the study of theology entirely after pursuing it in Leipzig as a student and switching to philosophy. He adored Christian Wolff and must have shared his concern about the still powerful role of theology faculties in the university.[79]

Yet the Halle Orphanage and its affiliated network laid a foundation for these later efforts. Those involved with this educational and scientific community trained thousands of skilled observers; they modeled new forms of scientific sociability, including how to investigate and even to reconcile a variety of perspectives; and they standardized a set of widely emulated procedures for teaching young people with objects and provided them with access to new technologies. In the process, they explored ways of linking the pursuit of knowledge in schools—through collaborative research and attention to observational techniques and methods of synthesis—with new forms of communal solidarity, expressions of friendship, camaraderie, and (if I may) love.

Acknowledgments

I will always be grateful to the individuals and institutions that made it possible for me to write this book. Peter Diehl, Amanda Eurich, Nancy Van Deusen, Chris Friday, and several other members of Western Washington University's history department were crucial early sources of encouragement. Without the help and support of Chris Friedrichs and Robert Brain at the University of British Columbia in Vancouver, this book probably would not have been written at all. Conversations with Mona Gleason, Anne Gorsuch, Joy Dixon, and many other wonderful colleagues in Vancouver also shaped it in countless ways.

As a Fulbrighter just out of college I taught in Halle's Orphanage complex and became fascinated with its history. The J. William Fulbright Program generously helped me return several years later by granting me the research fellowship that made my archival work possible. I received warm welcomes and good advice from Pia Schmidt, Udo Sträter, Christian Soboth, Thomas Müller-Bahlke, Britta Klosterberg, Holger Zaunstock, and Claus Veltmann. Anke Mies, Erika Papst, and Jürgen Gröschl taught me how to negotiate the Francke Foundation library and archive. Funding from the Fritz Thyssen Foundation allowed me to continue my work in Halle for several months. There I discussed my research in colloquia and informally with Renate Wilson, Jürgen Helm, Andreas Kleinert, Alexander Schunka, Jonathan Strom, Benjamin Marschke, Gunhild Berg, and Gita Rajan, who were all models of collegiality. Special thanks to Simon Grote, who helped me work out many of the issues I explore in this book through a series of conversations that originated in this archive. Simon Schaffer intervened at a key moment, inviting me to discuss the project with him and helping me think through what it might mean to see the Orphanage as a kind of scientific academy. In Cambridge, I was fortunate to have the chance to exchange ideas with Mirjam Brusius, Geoff Belknap, Alexander Wragge-Morley, and Chitra Ramalingam. Thanks to Ernst Hamm for his comments, to Matthew Jones for talking with me about Tschirnhaus at an early stage, and to Christopher Carlsmith and Christopher Corley for their engagement with a paper I presented on the Orphanage in Norrköping, Sweden.

I received additional funding from the Leibniz Institute for European History in Mainz and the Deutsche Akademische Austauschdienst; I thank Heinz Duchhardt, Irene Dingel, and several colleagues in Mainz for their support and interest in my work. My time at the Max Planck Institute for the History of Science in Berlin shaped my vision for this book in myriad ways. It also taught me a great deal about scientific community and collaboration. I am extremely grateful to Lorraine Daston and Hans-Jörg Rheinberger, who invited me to investigate Solomon's Temple models as part of the institute's Scientific Objects Research Network and to present my research in department colloquia. I received valuable feedback from colleagues at the institute, especially Tania Munz, Stefanie Klamm, Jeffrey Schwegman, Fabian Krämer, Fernando Vidal, Daniel Andersson, and Nicolas Langlitz. The project also benefited from conversations with Robyn Braun, Didier Debaise, and Monika Wulz, as well as Mary Terrall, Michael Gordin, Erika Milam, Alix Cooper, and many others. It was an honor to discuss work in progress with Claudia Stein and fellow participants in the Making of Early Modern Scientific Knowledge workshop at the University of Warwick and with Antonella Romano, László Kontler, Anthony La Vopa, Hans Erich Bödecker, and several other colleagues at the Central European University Summer School in Budapest. Thanks to James Delbourgo and Justin Smith for the opportunity to discuss Halle's curiosity cabinet at their "In Kind" workshop (supported by Cambridge University's Centre for Research in the Arts, Social Sciences and Humanities) and to Claudia Jarzebowski and Tom Safley for inviting me to discuss the Orphanage at the University of Pennsylvania. I am grateful for the feedback I received from those who participated in these events.

This book is better because of the constructive criticism and suggestions I have received at various stages from my editor Karen Merikangas Darling and from my external readers. I thank them heartily for all their good advice and the encouraging spirit in which it was offered. Some material in this book previously appeared in "Eclecticism and the Technologies of Discernment in Pietist Pedagogy," Journal of the History of

Ideas (2009): 545–567, and "The Model that Never Moved: The Case of a Virtual Memory Theory and Its Christian Philosophical Argument," Science in Context (2010). Thanks to Cambridge University Press and the University of Pennsylvania Press for allowing me to include some of this material here.

My colleagues and students at Sewanee have been allies in my quest to complete this book. I am especially grateful to Andrea Mansker and Julie Berebitsky for engaging me in focused conversations about the project and to Roger Levine and Nicholas Roberts for reading portions of my fifth chapter. A University Research Grant in the summer of 2012 allowed me to return to Germany so I could undertake a substantial revision of my third chapter. A grant from the Appalachian College Association has made it possible for me to channel my energies into seeing the book through its final stages. Thanks to several librarians and administrators at the University of Göttingen, the Herzog-August Bibliothek in Wolffenbüttel, the Staatsbibliothek in Berlin, the Franckesche Stiftungen Archiv und Bibliothek, and Halle's Stadtarchiv for their help with acquiring artwork and to Jean Ricketts for her help with organizing this artwork and the manuscript as well.

Finally, I would like to acknowledge the support of several friends and family members: special thanks to Sabine and Torsten Otto and the Wunder, Wunderlich, and Krauss families for showing me a side of Halle I wish more people could see. Even before I was serious about taking it on, Vaughn and Laurie Olthof insisted that a book about Halle's Orphanage was an excellent idea. Lisa Arakelian, Jill Eelkema, Julie and Raul Iribarren, Lois and Bob Franco, Sarah Vance, John and Dagmar and Kate Gundersen, and Kat and Charlie Zammit have been important sources of moral support, as have Richard and Marcia DeWaard, John and Kathy Whitmer, and David and Joanne Whitmer. Last, but not least, I thank Charles Whitmer, who has believed in this book for almost as long as I have.

Notes

CHAPTER 1

1. Kramer, *August Hermann Francke*, pt. 2, 328.
2. For more on pharmaceuticals see R. Wilson, *Pious Traders in Medicine*. For an overview of the mission see the essays in Gross, *Halle and the Beginning*.
3. Francke, "Projekt zu einem Seminario universali." All translations are my own, unless otherwise indicated in the bibliography.
4. Francke, "Was noch aufs künftige projectiret ist," 500.
5. Ibid., 500–501.
6. Ibid.
7. Francke, "Projekt des Collegii orientalis theologici," 282.
8. Jacobi, *"Man hatte von ihm gute Hoffnung,"* 16.
9. Müller-Bahlke, *Um Gott zu Ehren*, 5.
10. Beyreuther, *Geschichte des Pietismus*, 154.
11. Kramer, *August Hermann Francke*, pt. 2, 327.
12. Ibid., 328.
13. Ibid., 507. Kramer included an account of all the king's questions and Francke's answers during his visit as an attachment to this text. Ibid. 504–8.
14. Ibid., 508.
15. Ibid.
16. Freyer, "Verbesserte Methode des Pädagogii Regii," 420.
17. Kramer, *August Hermann Francke*, pt. 2, 329.
18. Ibid.
19. Ibid.
20. Ibid., 329–30.
21. Albrecht Ritschl influentially linked Lutheran Pietism to Reformed (Calvinist) piety movements in the Netherlands and Germany. See his *Geschichte des Pietismus*. Johannes Wall-

mann has strongly advocated a very narrow definition of Pietism, whereas
Martin Brecht and Hartmut Lehmann have championed a broader, more
inclusive way of understanding it. See Wallmann, "Was ist Pietismus?," "Eine
alternative Geschichte des Pietismus," "Pietismus—ein Empochenbegriff
oder ein typoloigscher Begriff?"; Brecht, "Probleme der Pietismusforschung";
and H. Lehmann, "Enger, weiterer und erweiterer Pietismusbegriff," and
"Erledigte und nichte erledigte." For accessible discussions of these debates in
English see Strom, "Problems and Promises of Pietism Research"; Ward, *The
Protestant Evangelical Awakening*; and Shantz, *An Introduction*.

22. Scholars of German Pietism have long distinguished "Halle Pietists" from
other Pietist groups. See Brecht, "August Hermann Francke und der Hallische
Pietismus." Radical Pietist groups included Philadelphians, the "Mother Eva"
society, "Communities of True Inspiration," and more. For an introduction
to radical Pietism in Germany see H. Schneider, *German Radical Pietism*; Breul
et al., *Der radikale Pietismus*; Shantz, *Between Sardis and Philadelphia*. For recent work on Moravian communities see Mettele, "Identities across Borders."
The "love feast" was an important ritual practice for many Pietist groups. See
Eller, "The Recovery of the Love Feast."

23. For more on Pietism studies and theologians in Germany today see Gierl,
"Im Netz der Theologen," 483–86. Thanks to Benjamin Marschke for recom-
mending this article.

24. Gierl, *Pietismus und Äufklärung*.

25. Religious enthusiasts claimed to enter into direct communication with God
through revelation; their critics considered this dangerous because demonic
forces might be behind the strange behaviors and abilities some exhibited,
including prophecy, trances, and spontaneous preaching. See Heyd, *"Be
Sober and Reasonable"*; and Johns, *The Nature of the Book*, 409ff.

26. Francke's personality and that of his mentor, Philipp Jakob Spener, have
been widely discussed in biographies, commemorative volumes, and sym-
posia. See Wendebourg, *Philipp Jakob Spener*; Wallmann, *Philipp Jakob Spener*;
Albrecht-Birkner, *Hoffnung besserer Zeiten*; Obst, *August Hermann Francke*; and
Kramer, *August Hermann Francke*, parts 1 and 2.

27. Stitziel, "God, the Devil, Medicine, and the Word."

28. See Lindberg, *The Pietist Theologians*; and Marschke, "Wir Hallenser."

29. Deppermann, *Der hallesche Pietismus und der preußische Staat*; Hinrichs,
Preußentum und Pietismus.

30. See Marschke, *Absolutely Pietist*, and "Lutheran Jesuits"; Fulbrook, *Piety and
Politics*"; Gawthrop, *Pietism and the Making of Eighteenth-Century Prussia*;
Melton, *Absolutism and the Eighteenth-Century Origins*; C. Clark, Iron King-
dom; Vierhaus, "The Prussian Bureaucracy Reconsidered"; and Raeff, *The
Well-Ordered Police State*. Crucial to these discussions has been the concept
of "confessionalization." See Zeeden, *Die Entstehung der Konfessionen*; and
Härter, "Sozialdisziplinierung." Heinz Schilling and Wolfgang Reinhard were
the most outspoken advocates of the confessionalization paradigm in Ger-

many. See Reinhard, "Zwang zur Konfessionalisierung?"; Schilling, *Religion, Political Culture*; and R. Hsia, *Social Discipline in the Reformation*.

31. LaVopa, *Grace, Talent, and Merit*, esp. 19–28. There was a long-standing tradition in Protestant Germany for families to provide "poor students"—also young boys attending Latin schools—with meals, even room and board, as an act of charity. By 1700, several universities had created large dining halls, where they offered "poor students" the chance to eat for free at the so-called *freie Tische*.
32. For further discussion of the Orphanage as a model of charity see R. Wilson, "Philanthropy in Eighteenth-Century Central Europe," and "Replication Reconsidered." For more on cameralism and the bureaucratization of daily life see Wakefield, *The Disordered Police State*.
33. See Jacobi's introduction to *"Man hatte von ihm gute Hoffnung"*; Müller-Bahlke, "Die frühe Verwaltungsstrukturen der Franckeschen Stiftungen"; and Sträter, "Pietismus und Sozialtätigkeit," 203ff.
34. Most German city orphanages were for children from honorable families; poor and abandoned children were not accepted in these institutions, especially if they were illegitimate. See Harrington, *The Unwanted Child*. Sometimes inspectors looked the other way and took in destitute, homeless children when no other alternative existed; this happened in Halle too. See Safley, *Charity and Economy*, and *Children of the Laboring Poor*; and Ulbricht, "Foundling Hospitals in Enlightenment Germany," 215.
35. W. Clark, *Academic Charisma*, 13.
36. Ibid.
37. For efforts to raise funds by selling medicines see R. Wilson, *Pious Traders*; and Quast, "Schwägerinnen."
38. Howard, *Protestant Theology and the Making*, 91–92. For more on the founding of the University of Halle see Holloran, "Professors of Enlightenment"; and Schrader, *Geschichte*.
39. Holloran, "Professors of Enlightenment," 3.
40. For the culture of disputation at German universities see W. Clark, *Academic Charisma*, 74–80.
41. Karant-Nunn, *The Reformation of Feeling*; Boros, De Dijn, and Moors, *The Concept of Love*; Gowing, Hunter, and Rubin, *Love, Friendship and Faith*.
42. Sulek, "On the Classical Meaning of Philanthropia," 386. Although the first Christians preferred the term *agape*, they used it interchangeably with the term *philanthropy* to refer both to the love of God for men and to the love of men for one another. See Downy, "Philanthropia in Religion and Statecraft," 200.
43. See AFST/H L 16: 4, 5, Früh-Predigten; Francke, *Schriftmäßige Betrachtung, La Philanthropie*, and *Philotheïa*.
44. Boros, De Dijn, and Moors, *The Concept of Love*, 81.
45. Ibid., 85. Leibniz was dissatisfied with Descartes's assertion that all matter, including the body, was passive, however. He developed his own doctrine of force that he described in the *Ipsa Natura*. See C. Wilson, "De Ispa Natura."

46. Jones, *The Good Life*; and Schneewind, "Philosophical Ideas of Charity."

47. Like Leibniz, Hoffmann considered himself a Cartesian, but he eventually argued against Leibniz's idea of a "pre-established harmony" between body and mind and referred instead to a "physical influx between the rational and the sensitive soul" (*De fato physic*, 1724). He also spoke out against Leibniz's doctrine of force, "defending a mechanical understanding of nature and a Cartesian account of animate and inanimate motion based on aether" (*De natura morborum*, 1699). Quoted in Naragon, "Friedrich Hoffmann," 528.

48. For more on Stahl see Geyer-Kordesch, *Pietismus, Medizin und Aufklärung in Preußen*; Reill, *Vitalizing Nature in the Enlightenment*; and King, "Stahl and Hoffmann."

49. I discuss these visits in chapter 2. For more on Tschirnhaus's visit see Ullmann, "August Hermann Francke"; and Müller-Bahlke, "Naturwissenschaften und Technik." For Leibniz's visits and correspondence with Francke see Utermöhlen, "Die Russlandthematik."

50. Sociologist of science Robert Merton argued against this tendency long ago. Although the "Merton thesis" is associated with a much older story about ascetic Protestantism and the origins of a "scientific revolution" that has been substantially and justifiably revised, it is still worth noting that Merton saw Pietists as proscience. Some, such as George Becker, disagreed with him on the grounds that "Pietism, unlike Puritanism, was decidedly an enemy of rationalism." Becker, "Pietism's Confrontation," 139–41. See also Merton's chapter "Puritanism, Pietism and Science" in *Social Theory and Social Structure* and his "Fallacy of the Latest Word."

51. Gay, *The Enlightenment*; Sheehan, "Enlightenment, Religion, and the Enigma of Secularization"; Coleman, "Resacralizing the World"; Sorkin, *The Religious Enlightenment*; Grote, "Review-Essay."

52. For an introduction to these discussions see Osler, "Mixing Metaphors," and *Reconfiguring the World*; and Ferngren, *Science and Religion*.

53. Outram, *The Enlightenment*, 2–3.

54. Many continue to treat Pietism and the Enlightenment as stable analytical categories in order to compare and contrast them. Albrecht Beutel, for example, describes them as "disputatious siblings" and "sometimes completely opposing, sometimes complimentary movements." See Beutel, "Causa Wolffiana," 159. For other examples of this tendency see Israel, *The Radical Enlightenment*, esp. 544ff; and W. Clark, *Academic Charisma*, 93–95.

55. Drechsler, "Christian Wolff"; Wolff, *Oratio*.

56. For a pathbreaking and now canonical study of a public controversy in the history of science see Shapin and Schaffer, *Leviathan and the Air-Pump*.

57. Jonathan Israel has advocated studying the Wolff episode in order to see "how structures of belief and sensibility in society interact dialectically with the evolution of philosophical ideas." *Enlightenment Contested*, 26. Yet as Robert Leventhal notes in a review of Israel's *Enlightenment Contested*, the resulting tendency has been to privilege "the intricacies of intellectual debate"

over any significant discussion of social reality. Leventhal, "Enlightenment Uncontested," 3.

58. Howard, *Protestant Theology and the Making.*
59. Herrnschmidt, "Von den rechten Grenzen," 14.
60. For examples of these assessments see Jacobi, *"Man hatte von ihm gute Hoffnung."*
61. Even Robert Boyle, who promoted an experimental program for natural philosophy, advocated a collaborative relationship between theology and natural philosophy. The Halle Orphanage published some of the earliest German translations of Boyle's *Seraphic Love* (1659–60), *Considerations Touching the Style of the Holy Scriptures* (1661), *Excellence of Theology compared with Natural Philosophy* (1664), and *Treatises on the High Veneration Man's Intellect Owes to God* (1684–85).
62. For more see Bruning, *Innovation in Forschung und Lehre,* 147–50.
63. In Kleinert, "Johann Joachim Lange," 82.
64. Wolff, *Anfangsgründe aller mathematischen Wissenschaften.*
65. In the wake of the expulsion, Wolff spoke out in favor of what he called his "freedom to philosophize." See Hettche, "On the Cusp," 92ff.
66. Wolff, *Vernünftige Gedanken.* Hettche notes that Halle's theologians must have realized "many of Wolff's philosophical views were consistent with the doctrine of pre-destination, a central tenet of Calvinist theology," which would have made them more inclined to see Wolff as a threat. Hettche, "On the Cusp," 93. Wolff produced a more complete version of his philosophical system (*Philosophia Rationalis*) in 1728; see Wolff, *Preliminary Discourse.*
67. Melton, *The Rise of the Public,* and "Pietism and the Public Sphere." Studies exploring links between urban sociability and networks of civic "natural history societies" in the second half of the eighteenth century include Phillips, *Acolytes of Nature*; and Lowood, *Patriotism, Profit, and the Promotion of Science.*
68. MacClellan, *Science Reorganized*; Lux, "Societies, Circles, Academies and Organizations"; Sutton, *Science for a Polite Society*; Biagioli, *Galileo, Courtier.*
69. See, for example, Shapin, *A Social History of Truth*; and Shapin and Shaffer, *Leviathan and the Air-Pump.*
70. Klein and Spary, *Materials and Expertise*; Stewart, "Other Centres of Calculation."
71. Foundational studies include Smith, *The Business of Alchemy*; and Nummendal, *Alchemy and Authority.*
72. See the classic study by Webster, *The Great Instauration*; Yates, *The Rosicrucian Enlightenment*; Eliav-Feldon, *Realistic Utopias*; Dickson, "Johann Valentin Andreae's Utopian Brotherhoods," and *The Tessera of Antilia*; Houston, "'Knowledge Shall Be Increased';" and Kanthak, *Der Akademiegedanke.*
73. Goodrick-Clarke, "The Rosicrucian Afterglow"; Murphy, *Comenius.*
74. Another vocal advocate of this kind of reform was Johann Joachim Becher. See Smith, *The Business of Alchemy,* esp. 84–86; Rossi, *Logic and the Art of Memory*; and Antognazza and Hotson, *Alsted and Leibniz.*

75. Ernst the Pious in Gotha was an early promoter of educational reform (see chapter 2); Germany's National Scientific Society, the *Leupoldina* (founded 1652), is a prime example of a new scientific society founded at this time.

76. "The Holy Roman Empire never possessed anything like the network of unofficial, local groups scattered through Italy, France, even Spain and provincial England." Evans, "Learned Societies in Germany," 130. See also Ornstein, *The Role of Scientific Societies*; Pedersen, "The Rise of the Academies," 480–88; and Stewart, *The Rise of Public Science*. See Florence Hsia's helpful introduction to the institutional history of early modern European science in "Mathematical Martyrs," 3–15.

77. Some have challenged the idea that scientific academies, not universities, were the preferred scientific communities in the seventeenth century. I agree with those who have pointed to the destructive effects of the Thirty Years' War on German universities in particular. See, for example, Evans, "Learned Societies." For a different view see Feingold, "Tradition versus Novelty"; and Gascoigne, "A Reappraisal."

78. See MacClellan, *Science Reorganized*, 68–74.

79. See Wakefield, *The Disordered Police State*, esp. 19ff.

80. Conrads, *Ritterakademien der Frühen Neuzeit*. A Ritter academy operated in Halle from 1685 to 1692, just before the Orphanage and university were founded.

81. AFST/W VII/I/20, Instruktionen für die Herumführer.

82. A few scholars have noticed a discernible "culture of empiricism" in Halle beginning around 1720, yet I think my findings show it was emerging earlier than this. See van Hoorn und Wübben, *"Allerhand nützliche Versuche."*

83. There has been a surge of interest in eclecticism by philosophers, however. See Gaukroger, *Francis Bacon*; U. Schneider, "Eclecticism"; and Albrecht, *Eklektik*.

84. For a helpful introduction to "practical mathematics" see Bennett, "The Challenge of Practical Mathematics." For mathematics and experimental philosophy see Shapin, "Robert Boyle and Mathematics."

85. There is a rapidly expanding literature on observation as a form of scientific experience. See Daston, "On Scientific Observation"; Daston and Lunbeck, *Histories*; Smith, *The Body of the Artisan*; Classen, *Worlds of Sense*; Schickore, *The Microscope and the Eye*; and Bleichmar, *Visible Empire*.

86. It was only in the nineteenth century that experiment came to be seen as more important than observation, which some claimed "could be safely left to untrained assistants." Daston and Lunbeck, introduction to *Histories*, 2–3. For instruments that signaled one's commitment to experiment and collective observation see Shapin, *The Scientific Revolution*, 95ff.

87. Daston, "The Empire of Observation," 87.

88. Roberts, "Situating Science in Global History"; Harris, "Confession-Building"; F. Hsia, *Sojourners*; H. Cook, *Matters of Exchange*.

89. Daston, "The Empire of Observation," 91–101.

90. I explore the interest of Halle-trained missionaries in names, ethnography, and observation in Whitmer, "What's in a Name?"
91. Sibum, "Experimentalists," 95; Roberts, Schaffer, and Dear, *The Mindful Hand*.
92. See Jones's discussion of Leibniz and "seeing all at once" in *The Good Life*, 229–66.

CHAPTER 2

1. Wollgast, *E. W. v. Tschirnhaus*, 21.
2. Mayer, "Am Rande der Gelehrtenrepublik," 26–27.
3. In Wollgast, *E. W. v. Tschirnhaus*, 41–42.
4. Maria Rosa Antognazza and Howard Hotson have argued for Leibniz's interest in radical Pietism and millennialism; see Antognazza and Hotson, *Alsted and Leibniz*; Shantz, "Conversion and Revival"; and Baumgart, "Leibniz und der Pietismus."
5. Hinrichs, *Preußentum und Pietismus*, 70. See also Müller-Bahlke's discussion in "Naturwissenschaft und Technik," 368.
6. See Tschirnhaus, "Brief an A. H. Francke," 318–19, and AFST/H C 263: 1, E. W. von Tschirnhaus to A. H. Francke, January 17, 1698.
7. Ibid.
8. Ibid.
9. Tschirnhaus, *Medicina Mentis*, and *Gründliche Anleitung*. There is also a manuscript version of the *Gründliche Anleitung* in the Orphanage's archive. See AFST/H D 85, 599–626.
10. Buchenau, *The Founding of Aesthetics*, 28.
11. Ibid., 29.
12. Ibid., 30.
13. Ibid.
14. As Jonathan Israel has explained, Spinoza was associated with a "widespread, formidably entrenched philosophical underground active in Holland, France, Britain, Italy and Germany" that pursued a "democratic republican agenda," including religious pluralism, toleration, and intellectual freedom. See Israel, *Radical Enlightenment*, 27, 31–32.
15. See Spener, "Brief an Ehrenfried Walther von Tschirnhaus in Kieslingswalde," 399–403, letter no. 95. Spener outlines three "scruples" he has with the Medicina Mentis in this letter, yet he refrains from attacking Tschirnhaus in the manner in which Christian Thomasius had done in his "Monatsgespräche" from 1688. See 399n2.
16. Sträter, "Zum Verhältnis," 88. Johann Jakob Spener spent at least a couple of weeks studying with Tschirnhaus, although the dates of his visit are not clear. See Spener, "Brief an Johann Jacob Spener in Leipzig," 136n28.
17. Tschirnhaus, *Gründliche Anleitung*, 8.
18. Ibid.
19. Ibid., 8–9.

20. He did this with his friend Paul Anton, who also received an appointment in Halle's theology faculty because of his connections to Spener. See Shantz, *An Introduction*, 103; and Hopf, "Anton, Paul."
21. Shantz, *An Introduction*, 107.
22. Shantz notes that university theology students began to sell their philosophy books, convinced they were no longer necessary. Ibid., 108.
23. For more on the Leipzig events see Gierl, *Pietismus und Aufklärung*.
24. Kevorkian, "Piety Confronts Politics," 145–46.
25. Baron Carl Hildebrand von Canstein was a soldier and court counselor in Berlin. He befriended Spener and used his network of patrons and contacts to raise funds for the Orphanage throughout Prussia and the Russian Empire. Canstein oversaw the delivery of funds, objects, and several instruments used in the schools, including a burning mirror donated by Tschirnhaus. He gave his entire fortune to Francke before he died, including several mines in the cities of Goddelsheim and Nordenbeck. See Schicketanz, *Carl Hildebrand*, and *Der Briefwechsel*. For Canstein's reference to Tschirnhaus's burning mirror see Müller-Bahlke, "Naturwissenschaft und Technik," 372n57.
26. Ullmann, "August Hermann Francke," 325–29.
27. See AFST/H D 99, Journal von Heinrich Julius Elers.
28. Ullmann, "August Hermann Francke," 332. The Prussian king offered him 3,000 taler and a position as president (*Rektor*) of the university in exchange for the recipe, but he refused. See Liebmann, "Walther von Tschirnhaus."
29. For several years, Tschirnhaus relied on Francke for financial support in exchange for help with planning the glassworks in Halle, but in the end, it was never built.
30. For mention of Tschirnhaus's visit to the Orphanage see Francke, "Brief an P. J. Spener," 380; Klüger, "Die pädagogischen Ansichten," 55; Francke, *Einrichtung des Pädagogii*, and *Ordnung und Lehrart wie selbige in dem Pädagogio*, 35.
31. Francke, *Ordnung und Lehrart wie selbige in dem Pädagogio*, 35.
32. Leibniz, "Brief an P. J. Spener," 703–4. Leibniz and Spener had already been corresponding extensively. See Antognazza and Hotson, *Alsted and Leibniz*, 137–38.
33. Leibniz, "Brief an P. J. Spener," 704.
34. Leibniz, "Brief an E. W. Tschirnhaus," 514–15.
35. Ibid., 515. "Solte man die Sachen in Schulen lehren, würde Schola recht Ludus werden."
36. Shapin, "Robert Boyle and Mathematics," 46–47.
37. Van Peursen, "E. W. von Tschirnhaus and the Ars Inveniendi," 402–3.
38. Ibid., 403–4. Discussed in detail in Tschirnhaus, *Gründliche Anleitung*, 8, 63–64.
39. Tschirnhaus, *Gründliche Anleitung*, 115. Points, lines, and curves are the only rational essences.
40. Ibid., 9.
41. Ibid., 114.

42. Ibid.
43. Ibid.
44. Tschirnhaus, "Brief an Leibniz," 464, letter no. 124.
45. These exercises resemble a program of practical mathematics developed in the Netherlands called "Duytsche Mathematique," designed to train young engineers and "fortificationists." Tschirnhaus very likely encountered it during his travels. For more on "Duytsche Mathematique" see Dijksterhuis, "Moving around the Ellipse," 92 and 107–8.
46. Van Peursen, "E. W. von Tschirnhaus and the Ars Inveniendi," 401n18.
47. Tschirnhaus, *Gründliche Anleitung*, 30.
48. For more discussion of Leibniz's interest in "easy and playful teaching of the sciences" see Jones, *The Good Life*, 254–55.
49. Weigel, "Kurtzer Entwurff"; Schlagenhauf, "Ansätze," 202–4; Meyer, *Leibniz*, 88–90.
50. See Ramati, "Harmony at a Distance"; Brather, *Leibniz und seine Akademie*; and chapter 3.
51. For an introduction to Campanella and Andreae see Headley, *Tommoso Campanella*; and Dickson, "Johann Valentin Andreae's Utopian Brotherhoods."
52. Leibniz, "Grundriss eines Bedenkens," 537.
53. Gerber, "Die Neu-Atlantis des Francis Bacon," 552–60.
54. See Bredekamp, "Leibniz's Theater of Nature and Art," and *The Lure of Antiquity*.
55. Dreyhaupt, *Ausführliche diplomatisch-historische Beschreibung*, 224.
56. There is a growing literature on the history of early modern collecting and the curiosity cabinet. For an introduction see Impey and MacGregor, *The Origins of Museums*; Findlen, *Possessing Nature*, and "Inventing Nature"; Bleichmar and Mancall, *Collecting across Cultures*; and Zytaruk, "Cabinets of Curiosities."
57. See K. Arnold, "Trade, Travel and Treasure," 267ff.
58. Francke, *Ordnung und Lehrart wie selbige in dem Pädagogoio zu Glaucha*, 73.
59. Ibid.
60. Michael Belk, a young man from London who studied in the *Pädagogium* and at the University of Halle from 1701 to 1708, noted in a letter he wrote Francke on his way home: "I have this morning bought of the steersman of our ship the ear of a whale, which with the first opportunity from England I intend to your curiosity chamber." AFST/H F 14: 289, Michael Belk to Francke, June 14, 1708. Many thanks to Lauren Lyons for alerting me to this passage.
61. Maria Sophie von Marschall sent a curious object along with a letter requesting the admittance of a boy named J. S. Geiersbach to the Orphanage. Stab/F 4b/13: 40, Maria Sophie von Marschall to A. H. Francke, January 25, 1699; Johann Hosmann von Rothenfels, writing from Vienna, sent a rare coin as thanks for admitting his son to the *Pädagogium*. Stab/F 27/13: 2, Johann Hosmann von Rothenfels to A. H. Francke, April 12, 1708.

62. Müller-Bahlke, *Die Wunderkammer*, 15.

63. Ibid., esp. 36, 54, 69; see also AFST/W XI/-/71, Inventarverzeichnisse der Kunst- und Naturalienkammer, and the first printed catalog, *Specification, derer Sachen* (1700).

64. Halle-trained missionaries consistently referred to the Orphanage's curiosity cabinet in their letters. See Whitmer, "What's in a Name?" 344–46; AFST/M 1 C 1: 27, Bartholomäus Ziegenbalg to A. H. Francke, October 1, 1706; AFST/M 1 C 1: 35, Bartholomäus Ziegenbalg and Heinrich Plütschau to Joachim Justus Breithaupt and A. H. Francke, October 17, 1707; D. Lehmann, *Alte Briefe aus Indien*, 42–46; and Stab/F 33/2: 2, Heinrich Plütschau to A. H. Francke, October 16, 1706. For more on the writing implements see Dolezel "Inszenierte Objekte," esp. 31–32; and Müller-Bahlke, *Die Wunderkammer*, 90–92.

65. For Jesuit "scientific missions" see F. Hsia, *Sojourners*; and Harris, "Confession-Building." For more on Leibniz's interest in a Protestant mission see Francis Merkel, "Missionary Attitude."

66. In Utermöhlen, "Die Russlandthematik."

67. Neubauer returned with drawings, notes and a report on Dutch relief institutions: see Neubauer, *Was bey Erbauung*; Stab/F 28/1: 2–4, Ergänzende Fragen Georg Heinrich Neubauers zum Studium der Waisenhäuser in Holland, and Aufzeichnungen; and Spaans, "Early Modern Orphanages."

68. Francke's "Historische Nachricht" was printed as an attachment to Spener's *Christliche Verpflegung der Armen*. See Hinrichs, *Preußentum und Pietismus*, 40–41.

69. Hinrichs, *Preußentum und Pietismus*, 41. See also "Projekte der ersten Jahre," in Brather, *Leibniz und seine Akademie*, 159–60n198; Franz Merkel, *G. W. von Leibniz und die China-Mission*, 190–201; and Collis, *The Petrine Instauration*, 410.

70. See Utermöhlen, "Die Russlandthematik," 110–11.

71. Leibniz, "Brief an Francke," 215–16, and AFST/H C 147: a, Gottfried Wilhelm Leibniz to A. H. Francke, August 7, 1697. Leibniz suggests he and Francke try to meet in Braunschweig. See also Utermöhlen, "Die Russlandthematik," 112.

72. Leibniz, "Brief an Francke," 216, and AFST/H C 147: a, Gottfried Wilhelm Leibniz to A. H. Francke, August 7, 1697.

73. In ibid. and in Utermöhlen, "Die Russlandthematik," 112.

74. Riley, "Leibniz's Political," 228.

75. See Jones, *The Good Life*; and Hadot, *Philosophy as a Way of Life*, and *What Is Ancient Philosophy?*

76. Riley, "Leibniz's Political," 229.

77. Ibid., 230.

78. See chapter 1 and Drechsler, "Christian Wolff."

79. Utermöhlen, "Die Russlandthematik," 112–13.

80. Ibid.

81. Ibid.

82. Hartkopf, *Die Berliner Akademie der Wissenschaften*; Vial, *Dr. Conrad Mel*.
83. See Mel, *Wäysen-Predigt*; and Stab/F 15, 1/5: 4, Conrad Mel to A. H. Francke. June 15, 1709. In this letter Mel provides Francke with a report on his own Orphanage in Hersfeld and his plans to send a student to observe the Halle facility.
84. See Stab/F 15, 1/5, esp. 10: Conrad Mel to A. H. Francke, January 2, 1711.
85. Mel, "Die Schauburg," and *Missionarius Evangelicus*.
86. Mel, "Die Schauburg," 231.
87. Ibid., 238–39. Other agents of the Orphanage abroad echoed this sentiment; see, for example, Anton Wilhelm Böhme's letter to Carl Hildebrand von Canstein, wherein he stresses the inroads Catholic missionaries have made due to their expertise in the "medical and mathematical sciences." AFST/H C 229: 58, Anton Wilhelm Böhme to Carl Hildebrand von Canstein, August 29, 1710.
88. Francke, "Projekt des Collegii orientalis theologici," 282.
89. Mel, "Die Schauburg," 233–34.
90. Leibniz spent time in Halle in 1707, 1711, and 1712; more evidence of meetings and correspondence between the two figures may exist. See Utermöhlen, "Die Russlandthematik," 117–18, especially n. 34.
91. Francke, "Brief an Gottfried Wilhelm Leibniz," 125–26.
92. Ibid., 126.
93. There was already a sizable community of German Lutherans there, who were involved in the area's mining industry.
94. Rescher, "Leibniz visits Vienna."
95. Leibniz, "Brief an A. H. Francke," 127, and AFST/H C 147: f, Gottfried Wilhelm Leibniz to A. H. Francke, January 17, 1714.
96. Ibid. 127.
97. Francke, "Brief an Gottfried Wilhelm Leibniz," 128.

1. Bennett and Mandelbrote, *The Garden*, esp. 107–9. For an introduction to his life and projects see Murphy, *Comenius*.
2. After reading the *Door of Language Unlocked*, Hartlib invited Comenius to London to discuss his ideas with a circle of similarly minded reformers. Comenius arrived in 1641 and worked with the Hartlib circle to produce his best-known work, the *Orbis sensualium Pictus* (1658). See Bennett and Mandelbrote, *The Garden*, 108; Schadel, *Sehendes Herz*, 23–24; and Mahnke, "Der Barock Universalismus des Comenius," esp. 111ff. For more on Hartlib and Comenius see Webster, *The Great Instauration*; Turnbull, *Hartlib, Dury and Comenius*; and Greengrass, Leslie, and Raylor, *Samuel Hartlib*. There were communities of brethren in several Polish and Czech cities, especially in Lissa, but church leadership operated underground out of fear of persecution. The brethren saw Jan Hus as the founder of their movement; he spoke out against

indulgences and the celibacy of priests in Bohemian and Moravian lands before the Protestant Reformation.

3. Comenius, *Naturall philosophie*. Efforts to create a pious natural philosophy had been ongoing since the late Renaissance, especially in light of a still widely held conviction that Aristotelian physics was too dominated by "heathen" ideas about the natural world. See Blair, "Mosaic Physics."

4. See Hinrichs, *Preußentum und Pietismus*, 37ff; Cervenka, *Die Naturphilosophie des Johann Amos Comenius*, 144, 164–65.

5. Blair, "Mosaic Physics," 32: "I conclude that what they shared most effectively was an agenda rather than a practice."

6. In 1723, Johann Ernst Bessler described the Halle Orphanage as a place where the children of Protestants attend "alongside Papists" and "do not hear even the slightest hint of confessional strife." Bessler, *Der Rechtglaubige Orffyreer*, 8–9. In this same passage, he describes plans to start a similar institution in Carolshaven (i.e., Bad Karlshafen). Bessler, a.k.a. Orffyreus, is best known for his construction of a perpetual motion machine in Kassel. For more see Schaffer, "The Show That Never Ends."

7. See Antognazza, *Leibniz*, 398ff; and Schunka, "Daniel Ernst Jablonski," 30–31 and 38–40.

8. The revision reveals an interest in "natural theology" or "physico-theology," although this phrase was generally not used in Halle. For more on "physico-theology" and its relationship to history and ethics see Olgilvie, "Natural History, Ethics, and Physico-Theology"; Vidal, "Introduction: Knowledge, Belief"; and Trepp, *Von der Glückseligkeit*.

9. For more on arguments for and against religious toleration in early modern Europe see Marshall, *John Locke, Toleration*. For the impact of Calixt and irenical theology on Leibniz see Antognazza, *Leibniz*, 46–50.

10. Wallmann, "Union, Reunion, Toleranz," 27.

11. Mager, *Georg Calixts theologische Ethik*; and Schlüssler, *Georg Calixt*.

12. Jung, *Das Ganze der Heiligen Schrift*.

13. Ibid., 35–37.

14. Sträter, *Sonthom, Bayly, Dyke and Hall*.

15. Sträter, *Meditation und Kirchenreform*, and "Wie bringen wir den Kopf in das Herz?"

16. Comenius studied in Heidelberg from 1613 to 1614 at precisely the same moment Pareus published his *Irenicum*. See Hotson, "Irenicism and Dogmatics," and "Irenicism in the Confessional Age"; and Nischan, "John Bergius."

17. See Gorsky, *The Disciplinary Revolution*, esp. 79–98.

18. Grau, *Die Preußische Akademie*, 58–59.

19. Lacroze was a former Benedictine monk who fled Paris and converted to Calvinism upon arriving in Berlin in 1696; he began corresponding with Leibniz in 1704 and with Francke shortly thereafter. See Brather, *Leibniz*, 320–23; Mulsow, *Die drei Ringe*; Pott, Mulsow, and Danneberg, *The Berlin Refuge*; and AFST/M 2 A 5, A. H. Francke, G. A. Francke and Maturin Veyssière La Croze.

20. For more on Leibniz's efforts to bring about confessional reconciliation see Jordan, *The Reunion*; Fischer "Leibnizens kirchenpolitische Wirksamkeit"; and Antognazza, *Leibniz*, esp. 398–406.

21. The Sacra Congregation de Propaganda Fide was a Catholic organization that promoted the cause of conversion through multilingual educational institutions and publications. See Brunner, *Halle Pietists in England*, 23. For more on moral reform efforts and philanthropic institutions in England at this time see Davison et al., *Stilling the Grumbling Hive*; and Cunningham and Ines, *Charity, Philanthropy, and Reform*.

22. Schunka, "Daniel Ernst Jablonski," 23–41, and "Zwischen Kontingenz." Jablonski also suggested using the Anglican liturgy and the English Book of Common Prayer, which he translated into German, to unite the Lutheran and Reformed churches in Prussian lands.

23. For more on Slare see M. Hall, "Slare, F. R. S. (1648–1727)"; and Golinski, "A Noble Spectacle."

24. Leibniz and Molanus, "De Unione Protestantium." For more on Molanus see Masser, *Christóbal*.

25. Jordan, *The Reunion*, 200.

26. Ibid., 201–2.

27. Ibid., 204.

28. Antognazza, *Leibniz*, 400.

29. Quoted in ibid., 401.

30. Christian Thomasius was a follower of Pufendorf and developed his own public law course in Halle grounded in the tenets of Pufendorf's political philosophy. See Eiskildsen, "Christian Thomasius," 323; and Brockliss, "Curricula: The Faculty of Law," 602.

31. I. Hunter, "The Love of a Sage," 170–71; see also Haakonssen, *Natural Law and Moral Philosophy*, esp. 46–49.

32. I. Hunter, "The Love of a Sage," 170–71.

33. Friedrich Wilhelm I cultivated a reputation as the "soldier king." He has a much harsher, more conservative reputation than his father, although recent scholarship suggests the emphasis on his harshness may have been exaggerated after his son, Friedrich "the Great" took the throne in 1740. See Friedrich, *Brandenburg-Prussia*, 99.

34. Böhme, *Das Sechste Büchlein*.

35. For "Anschauung" as a concept in philosophy see Kaulbach, "Anschauung"; and Kohlenberg, "Anschauung Gottes."

36. Hugo's *Pia Desideria* was published in Gotha in 1707. First published in Antwerp in 1624, it was translated into Spanish in 1626; German, French, and English in 1627; Dutch in 1629; and Italian in 1633. See Reimbold, *Pia Desideria*, 8–9; and Camille, "Before the Gaze."

37. Kühnel, *Comenius und der Anschauungsunterricht*, 22.

38. Ibid., 22–23.

39. Arndt, *Wahres Christentum*, 29.

40. Spener, "Pia Desideria."
41. Lewis Bayly's treatise *The Practice of Piety* was widely circulated. It was first translated and printed in German in 1631 and went through several editions. Comenius translated it into Czech as early as 1633. See Sträter, *Sonthom, Bayly, Dyke and Hall.* For more on Spener and his famous question "How do we bring the head into the heart?" see Lindberg, *The Pietist Theologians,* 8ff; and Sträter, "Wie bringen wir den Kopf in das Herz?"
42. Spener, *Theologische Bedenken,* esp. 28 and 132–33.
43. Spener, "Pia Desideria," 101–2.
44. For the prominence of emblems and symbolic imagery in Jesuit colleges see van Vaeck and Mannings, *The Jesuits and the Emblem Tradition;* and Chorpenning, *Emblemata Sacra.*
45. G. Arnold, *Göttliche Liebes-Funcken.*
46. S. Clark, *Vanities of the Eye,*161–203; Scribner, "Ways of Seeing in the Age of Dürer," 102–3.
47. Francke, *Lectiones paraeneticae,* 130–31.
48. Francke, "Die Lehre von der Erleuchtung."
49. Ibid. See also AFST/H L 16: 4, 5, Francke, Früh-Predigt; Francke, *Schriftmäßige Betrachtung, La Philanthropie,* and *Philotheïa: Oder die Liebe zu Gott.*
50. Boyle, *Seraphic Love.* For more on Boyle see M. Hunter, *Boyle.*
51. Boyle, *Seraphic Love,* 8/11 and 13/15.
52. See Boyle, *Amor Seraphicus.* A second edition appeared in 1709 along with German translations of Boyle's *Considerations Touching the Style of the Holy Scriptures, Excellence of Theology compared with Natural Philosophy,* and *Treatises on the High Veneration Man's Intellect Owes to God.* See Boyle, *Auserlesene Theologische Schrifften.* The community also translated Boyle's *Considerations about the Reconcileableness of Reason and Religion* but did not publish it; see AFST/B 50: 00, Einige Betrachtungen.
53. See Stab/F 30/47: 1-10, Frederick Slare to A. H. Francke, ca. 1704–22; Schunka, "Zwischen Kontingenz," esp. 89–91; and Brunner, *Halle Pietists in England,* esp. 81–83. One of the boys, Henry Hastings, was rumored to have developed his own perpetual motion machine while living in the Orphanage. See Brunner, *Halle Pietists in England,* 91n147.
54. Francke, *Das Auge des Glaubens,* and *Die Klugheit der Kinder des Lichts,* 23. He preached this sermon for the first time on August 13, 1713.
55. Sabunde, *(Oculus Fidei) Theologia Naturalis.*
56. See Boros, De Dijn, and Moors, *The Concept of Love;* and Vidal, *The Sciences of the Soul.*
57. See Ahnert, *Religion and the Origins,* esp. 107, 116; and Thomasius, *Versuch.*
58. Ahnert, *Religion and the Origins,* 101.
59. Ibid., 103.
60. Ibid., 30.
61. Quoted in Riley, *Leibniz's Universal Jurisprudence,* 156–57.
62. Ahnert, *Religion and the Origins,* 17.

63. Quoted in ibid., 17.
64. Judd Stitziel has noted that both Spener and Francke wanted to better understand the cause of the women's prophetic behavior; however, they also "took up defensive positions, privately worrying about the public and political consequences of their connections to the women." Stitziel, "God, the Devil, Medicine, and the Word," 317.
65. Bahr, *Höchstverderbliche Auferziehung*, 21.
66. Ibid., 23.
67. Sturm said he was simply following the recommendations of the ancient Stoics and the Dutch humanist Justus Lipsius, who stressed the importance of election or personal choice in philosophy. Lipsius's *Manductio ad stoicam philosophiam* (1604) stated that "a method of critical choosing or election should have guided philosophers all along and helped them avoid sectarian disputes or dogmatic errors." Quoted in Mercer, "Platonism and Philosophical Humanism," 26. See also Kelley, "Eclecticism and the History of Ideas," 580; Albrecht, *Eklektik*, 318–19, and "Hypothesen und Phänomene," 121; and Whitmer, "Eclecticism and the Technologies."
68. Herrnschmidt studied at the University of Altdorf from 1696 to 1698, after which he moved to Halle and worked closely with Francke and his colleagues in the theology faculty, Joachim Justus Breithaupt and Paul Anton. He returned home to Bopfingen in Schwaben around 1702, where he married and assisted his father, who was also a preacher. He was invited to return to Halle as a professor of theology in 1715 and was given the subdirectorship of the Orphanage as part of this promotion. "Johann Daniel Herrnschmidt," 221–22.
69. Herrnschmidt, "Von den rechten Grenzen," 11–12.
70. "Einer alleine kan nicht alles ausforschen; darum soll man sich die Wahl frey behalten," In Herrnschmidt, "Von den rechten Grenzen," 9.
71. Ibid.,11.
72. Ibid., 11–12.
73. Ibid., 17–18.
74. Ibid.
75. Ibid.
76. Wolff, *Vernünftige Gedanken*.
77. Ibid., 377.
78. Francke visited Buddeus in Jena around the time that tensions were mounting in Halle; Buddeus also hosted Francke's son Gotthilf while he was studying in Jena. See Stab/F 7/14: 37 and 38, Johann Franz Buddeus to A. H. Francke, February 2 and 16, 1719.
79. As Timothy Hochstrasser has explained, Buddeus distinguished between moral philosophy and moral theology: "*Philosophia moralis* takes its stand on reason, whereas *theologia moralis* takes revelation as its source of authority. It is moral theology that is decisive for the Christian, and therefore moral philosophy can only be a handmaid in the task of working out the correct

NOTES TO PAGES 53–57

pattern of behavior owed by the individual Christian man or woman to God and his or her neighbor. Wolff's crime in the eyes of Buddeus was exactly that of having reversed these priorities." Hochstrasser, *Natural Law Theories*, 150.

80. Quoted in ibid.,153. See also Buddeus, *Elementa philosophiae practicae*.

81. Hochstrasser, *Natural Law Theories*, 151.

82. Ibid., 153.

83. See Blair, "Mosaic Physics," esp. 37n10; and Buddeus, *Introductio ad historiam philosophiae ebraeorum*.

84. Rappaport, *When Geologists Were Historians*, 65–66.

85. Cervenka, *Die Naturphilosophie*, 172.

86. Buddeus listed this as one of his main concerns in *Bedencken über die Wolffianische Philosophie*, esp. 56–57. Despite efforts to defend his idea that "perfection can be maximized in a world that has a beginning but no end," Leibniz's system of metaphysics, including his cosmogony, remained associated with this idea. Futch, "Leibniz on Plenitude," 541.

87. Cervenka, *Die Naturphilosophie*, 174–75.

88. Ibid.

89. Ibid., 213–42.

90. Comenius did not completely disregard mathematics and logic, but like many advocates of experimental philosophy, he preferred practical over higher mathematics.

91. Bennett and Mandelbrote, *The Garden*, 109.

92. Francke, *Ordnung und Lehrart*; see also Peschke, "Die Reformideen des Comenius."

93. Reyher, *Kurtzer Unterricht*. See Albrecht-Birkner, *Reformation des Lebens*, 439n45; and Michael, *Die Welt als Schule*.

94. See Schunka, "Daniel Ernst Jablonski," 38–39. Schunka notes that Francke asked Jablonski to do the translations but Jablonski lamented he had neither the time nor sufficient knowledge of Czech to take this on.

95. Comenius, *De Rerum Humanarum Emendatione Consultatio Catholica*, and *Historia Fratrum Bohemorum*. For many years these original manuscripts were believed to be lost, but they were found in the Orphanage archives and returned to Prague relatively recently. see Hofmann, "Consultatio Catholica de emendatione rerum humanarum," and "Die 'Halleschen Funde.'"

96. Francke, "Historische Nachricht," 43–44.

97. Francke, *Kurzer und einfältiger Unterricht*, 37.

98. Ibid., 32.

99. Francke, *Ordnung und Lehrart*, 64.

100. Ibid.

101. Ibid., 73.

102. Hochstrasser, *Natural Law Theories*, 152n5; Buddeus, *Elementa philosophiae instrumentalis*, and *Elementa philosophiae theoreticae*.

103. Schmidt-Biggemann, "Die Historisierung der 'Philosophia Hebraeorum,'" esp. 120–22.

104. Freyer, *Verbesserte Methode*, 116–18.
105. Ibid., 118. Herrnschmidt's "Von den rechten Grenzen" is mentioned here as well.
106. Ibid., 122–28.
107. Ibid., 128–33.
108. Ibid., 134–40.
109. Ibid., 82. Freyer lists Wolff's *Beginner's Guide* as a resource for teachers directing experiments during recreational periods as well, particularly his sections on hydrostatics, aerometry, and hydraulics. See ibid., 133.
110. Ibid., 82.
111. Ibid., 87.
112. Ibid.
113. Ibid., 70.
114. Ibid., 98.
115. Ibid.
116. Howard, *Protestant Theology and the Making*, 97.
117. Although increasingly concerned about it by the 1720s, Halle's theologians did not entirely oppose the study of philosophy, as some scholars have argued (see Sparn, "Philosophie"). In "Project des Collegii orientalis theologici" Franke recommended prospective students pursue philosophy: "They should also be careful to demonstrate their serious study of philosophy . . . it would also be nice if they endeavor to apply what they learn in physics, mathematics and philosophy as well." Francke, "Projekt des Collegii orientalis theologici," 282.
118. Wolff, *Eigene Lebensbeschreibung*, 190. In 1739, Friedrich Christian Baumeister wrote a short biography of Wolff, called *Vita, fata et scripta Christiani Wolfii philosophi* (Leipzig: Richter, 1739). He asked Wolff to make additions and corrections to it and planned to publish a second edition using Wolff's notes, which never appeared. In 1841, Heinrich Wuttke used Wolff's notes and the Baumeister text to produce the *Eigene Lebensbeschreibung*. It is "autobiographical" in that it contains sections written by Wolff (in the first person), but it is very carefully constructed.
119. Ibid., 190. Wolff notes there were still lingering tensions between the faculties of law and theology in Halle and that Thomasius was particularly concerned that his philosophy was not being taken seriously. See also Wolff, *Biographie*.

CHAPTER 4

1. See de Chadarevian and Hopwood, *Models*; Evers, *Architekturmodelle der Renaissance*; Hughes, "Model Builders"; D. Baird, *Thing Knowledge*; Morgan and Morrison, *Models as Mediators*; and Mosley, "Objects of Knowledge."
2. Schaffer, "Machine Philosophy," 157–58; Warner, "What Is a Scientific Instrument?," 83.

3. Rich, "Representing Euclid."
4. Sturm, *Kurtze Vorstellung*, 3–4. For Sturm as a Berlin Academy member see Brather, *Leibniz und seine Akademie*, esp. 358.
5. Hamblin and Seely, *Solomon's Temple*; von Naredi-Rainer, *Salomos Tempel und das Abendland*.
6. Biesler, *BauKunstKritik*, 44; Rykwert, *On Adam's House in Paradise*.
7. See "Temple," 377.
8. Biesler, *BauKunstKritik*, 46; Sturm, *De sciagraphia templi Hierosolymitani*, and *Nicholas Goldmanns vollständige Anweisung zu der Civilbaukunst*.
9. See Lawlor, *Sacred Geometry*; Scott, *Architecture for the Shroud*, specifically the chapters "Faith in Geometry" and "Imagery of Devotion and Dynasty."
10. Schaper, "The Jerusalem Temple."
11. W. Lynch, *Solomon's Child*; Gerber, "Die Neu-Atlantis des Francis Bacon"; McClellan, *Science Reorganized*, 49–50.
12. See Jones, *The Good Life*; and Daston, "On Scientific Observation," 97.
13. Harries, *The Ethical Function of Architecture*, esp. 103–10; Summers, *Real Spaces*, 204.
14. Semler, *Nützliche Vorschläge*, 10.
15. In S. Lange, *Leben Georg Friedrich Meiers*, 20–21. Meier became a member of the Berlin Academy in 1751. For more on Meier see Pozzo, "Prejudices and Horizons," esp. 189ff.
16. S. Lange, *Leben Georg Friedrich Meiers*, 20–21.
17. Ibid.
18. See AFST/H A 173: 1, Tagebuch von A. H. Francke. Semler and Francke were both adjunct pastors of Halle's Saint Ulrich church and part of a royal commission formed in 1720 to investigate allegations of corrupt handling of the church's "widow's fund"; a letter to King Friedrich Wilhelm contains a report from this commission that refers to the archdeacon Francke as Semler's "good friend." GSPK, I HA Rep 52 Nr 159b, February 5, 1721.
19. C. H. von Canstein and A. H. Francke's exchange regarding Semler is discussed in Müller-Bahlke, "Der Realienunterricht," 59. See also Schicketanz, *Der Briefwechsel*.
20. Gaukroger, "Descartes' Early Doctrine"; Jones, *The Good Life*, 64–65.
21. Semler, *Nützliche Vorschläge*, 9–10. I have translated the German word *Erkenntnis* as "understanding" and *Gemüth* as "character"; however, there are really no equivalent words in English.
22. Semler, *Der Tempel Salomonis*, 1.
23. Rosenau, *Vision of the Temple*; Bennett and Mandelbrote, *The Garden*; Sheehan, "Temple and Tabernacle"; Carpo, "How Do You Imitate a Building That You Have Never Seen?" For an introduction to skenographia and ichnographia see Summers, *Real Spaces*, 497–89.
24. Offenburg, "Jacob Jehuda Leon (1602–1675)"; Robinson and Adams, *The Diary of Robert Hooke, 1672–1680*, 179.
25. See Vogelsang, "Archaische Utopien," 25–27; Uffenbach, *Merkwüridge Reisen*,

115–17. This temple model was displayed in Dresden for much of the eighteenth century. It is currently on display in the city museum of Hamburg.

26. Sheehan, "Temple and Tabernacle," 254.
27. Ibid., 259.
28. S. Lange, *Leben Georg Friedrich Meiers*, 20–21.
29. Sturm, *De Philosophia sectarian et elective*.
30. See Mercer, "Platonism and Philosophical Humanism," 33–35.
31. Semler, *Der Tempel Salomonis*, 8.
32. Ibid.
33. Busch, "Johan Lund," 62–67; Vogelsang, "Archaische Utopien," 297; Lund, *Die Alten Jüdischen Heiligthümer*.
34. Semler, *Der Tempel Salmonis*, 7.
35. Ibid., 6.
36. Ibid.
37. Ibid.
38. AFST/H E 61, Manuskript für den Realienunterricht, 557.
39. Wolff, *Mathematisches Lexicon*, 502ff.
40. Lange, *Leben Georg Friedrich Meiers*, 20–21.
41. Semler, *Der Tempel Salmonis*, 6.
42. Ibid., 80–81.
43. Francke, "Brief an Friedrich II," 326–27.
44. Francke, "Kurzer und einfältiger Unterricht," 108.
45. Ibid., 108; Boecler, *Characteres*; J. Hall, *Characteres der Menschen*.
46. See Bos, "Individuality and Inwardness," 151.
47. Ibid., 143.
48. J. Hall, *Characteres der Menschen*, 4.
49. Ibid.
50. Ibid., 10.
51. See M. Cohen, "La Bruyère and the 'Usage' of Childhood." Claude Fleury was famous for tutoring the Dukes of Anjou, Burgundy, and Berry.
52. Francke, "Kurzer und einfältiger Unterricht," 109. The Orphanage published Gottfried Arnold's *Vitae Patrum* (Lives of the Old Testament Fathers) in an effort to realize this goal.
53. Quoted in Menck, *Die Erziehung der Jugend*, 44–47.
54. In Francke, "Kurzer und einfältiger Unterricht," 135.
55. Johns, *The Nature of the Book*, 409.
56. See Whitmer, "The Model That Never Moved," esp. 293–302, and "Unmittelbare Erkenntnis."
57. *Hallesche Anzeigen* (February 9, 1739), 86–87. Several scholars have discussed the importance of "virtual witnessing," defined as "the pictorial and verbal means through which readers are given a sense of having experienced the events described in a report"—for example, a description of an experiment. M. Lynch, "The Production of Scientific Images," 30; Shapin and Schaffer, *Leviathan and the Air-Pump*.

58. Freyer, "Verbesserte Methode," 419.
59. Ibid., 123.
60. Ibid.
61. Ibid., 124.
62. Ibid., 125.
63. Ibid.
64. Ibid.
65. Lange, *Leben Georg Friedrich Meiers*, 20–21.
66. See Müller-Bahlke, "Der Realienunterricht," 58; and Whitmer, "Eclecticism and the Technologies," esp. 563–65.
67. AFST/H E 61, "Astronomia mechanica oder eigentlichen Modell das grossen Welt Gebaudes nach dem Systemate Copernicano und Tychanico," in Manuskript für den Realienunterricht, esp. 331, 739–63, 829.
68. Ibid., 859, 864–65.
69. Ibid., 739–40.
70. Ibid., 740.
71. Ibid.
72. Ibid., 886.
73. Ibid., 896–97.
74. Ibid., 897–901.
75. *Gründlicher Unterricht von der Graphice* [1717].
76. Ibid., 426–27.
77. Ibid.
78. AFST/W XI/-/72, Demonstration des Modells der Stiftshütte Mosis; Semler, *Eigentliches Modell und materielle Figur zu Erläuterung sehr vieler Örter Heiliger schrift aufgerichtet*; Semler, *Die Stadt Jerusalem*, and *Palaestina*.
79. For more on radical religious communities and their interest in sacred geometry and architecture see Bernet, *"Gebaute Apokalypse.'"*
80. For Francke's use of the phrase "prophet school" see his "Project zu einem Seminario Universali."
81. Dilherr, *Propheten Schul*, 2–3.

CHAPTER 5

1. Leibniz, "Concept einer Denkschrift," 247.
2. Ibid.
3. The Board of Longitude convened after the passing of the Longitude Act of 1714 and promised sizable rewards to individuals who could come up with a reliable method for measuring longitude. For an accessible introduction to the longitude problem see Sobel, *Longitude*; and Andrews, *The Quest for Longitude*.
4. See W. Clark, *Academic Charisma*, esp. 374–75.
5. Heinrich W. Ludolf published the first grammar of the Russian language, the *Grammatica Russica*, in 1696. He had familial connections to the Danish crown and spent several years in London and in Russia. His uncle, Hiob, lived

in Halle from November 1696 to March 1698 and was an early member of the Berlin Academy. Heinrich stayed with Francke in Halle from the end of 1697 to April 1698 and corresponded with Leibniz during this time. He was also present during Tschirnhaus's visit to the community. See R. Wilson, "Heinrich Wilhelm Ludolf," esp. 88ff; and Cracraft, *The Petrine Revolution*, 35–38. For some clues regarding Eberhard's movements during this period see Helm and Quast, *Fromme Unternehmer*, 173; and Jakubowski-Tiessen, "Der Pietismus in Dänemark," 446.

6. AFST/H C 265: 3, C. Eberhard to A. H. Francke, March 20, 1706. Eberhard noted that then there would then be a total of three exemplary orphanages and *Pädagogia*, including one funded by the Duke of Hesse, who supported the efforts of Conrad Mel to develop an orphanage modeled after Halle's in Hersfeld.

7. Jakubowski-Tiessen, "Der Pietismus in Dänemark," 468.

8. See Helm and Quast, *Fromme Unternehmer*, 173.

9. Johann Tribbechow became a member of the Collegium Orientale in 1702; he was a professor of philosophy at the university and served as an adjunct in the theology faculty in 1705 before leaving for London. He obtained this post through Heinrich W. Ludolf's interventions.

10. AFST/H C 265: 7, C. Eberhard to Francke, April 25, 1710; see also Stab/F 22/5: 41, Gottfried Vockerodt to A. H. Francke, December 17, 1710.

11. Quoted in Brunner, *Halle Pietists in England*, 45. See also McClure, *A Chapter in English Church History*, esp. 24ff.

12. For an accessible introduction to these conflicts see Frost, *The Northern Wars*.

13. Winter, *Halle als Ausgangspunkt*, 73.

14. Ibid., 76–77. When he first arrived in Moscow, Scharschmidt was shunned by the German Lutheran congregation there, likely because of his ties to Spener and Francke. He was taken in initially by Laurentius Blumentrost (he later became Tsar Peter's personal physician), until he had learned some Russian and become confident enough to forge new connections in Narva and Astrachan.

15. The tsar visited the ironworks and on one occasion stayed for an entire month. Collis, *The Petrine Instauration*, 344.

16. Müller studied theology in Halle and then spent several years in the Netherlands and Paris, eventually returning to Moscow to take over his family's ironworks. He stayed in regular contact with friends in Halle: for example, in April 1710, he wrote to Francke from Paris, sending him information about François Fenelon as well as his impressions of Louis XIV after meeting him during a personal audience. See AFST/H A 183: 9, Peter Müller to A. H. Francke, April 5, 1710. Eberhard married Müller's sister, Anna, in Hamburg in 1719.

17. Leibniz's plans for the Berlin Academy and a system of academies, along with Halle's Orphanage and university, served as blueprints for these efforts. See Lipski, "The Foundation of the Russian Academy of Sciences," 349–54; Gordin, "Importation," 2; and O. Baird, "I Want the People to Observe and to Learn!"

18. Winter, *Halle als Ausgangspunkt*, 84.
19. Ibid., 85. When Bruce, Cruys, von Weyde, and Tsar Peter were in England in 1698, they met with members of the Royal Society, including Isaac Newton and Edmund Halley, whose magnetic theory would later make a major impression on Eberhard. It is not clear if the group actually visited the Royal Society or saw any experiments performed; however, they did visit its museum. See Gordin, "Importation," 4. Gordin notes that in 1717, Tsar Peter traveled to Paris and visited the French Academy of Sciences, which likely also served as a model for the Saint Petersburg Academy. Bruce remained behind in England after the tsar left in order to study with individual members of the Royal Society, purchase scientific instruments, and recruit skilled individuals to come back with him to Russia. See Cross, *Peter the Great*, 31.
20. Winter, *Halle als Ausgangspunkt*, 83. See Collis, *The Petrine Instauration*, esp. 47–119.
21. Winter, *Halle als Ausgangspunkt*, 114.
22. Mühlpfordt, "Halle—Russland," 64–65; Wotschke, "Das pietistische Halle," 475.
23. A variety of Orphanage Observators in Russia were involved in these plans, including Scharschmidt, Ulrich Thomas Roloff, Curt Friedrich von Wreech, and Daniel Gottlieb Messerschmidt. See AFST/H C 491, Projektentwurf zur Gründung eines Waisenhauses in Tobolsk. For recent work on Messerschmidt and his collecting practices see te Heesen, "Boxes in Nature."
24. C. Eberhard, *Der innere und äussere Zustand*.
25. Ibid., 22–23.
26. Ibid.
27. Russia was already a supplier of magnetic iron ore; there were a wide variety of Russian lodestones—often placed in elaborate ornamental cases—in circulation across Europe. See Ryan, "Scientific Instruments in Russia," esp. 376–77.
28. C. Eberhard, *Versuch*, 9.
29. Ibid.
30. For the magnet as a "model of cosmic forces" see Pumfrey, "Mechanizing Magnetism," and "Magnetical Philosophy."
31. Bennett, "The Challenge of Practical Mathematics," 188.
32. Baldwin, "Magnetism."
33. See Pumfrey, *Latitude*; and Gilbert, *On the Magnet*. He argued that all magnets are impelled by an immaterial force to rotate until they achieve a "harmonious position" or "magnetic union" and that they always align themselves in fixed directions. Therefore, the magnetized earth had to revolve, and it had to revolve in only one direction.
34. In his book *The Magnet*, Athanasius Kircher described an *Instrumentum Pantometrum, Ichnographicum Magneticum*, or a device that allowed one to measure everything. It was basically a "measuring table" with a built-in compass. See Vollrath, "Das Pantometrum Kircherianum." Kircher probably relied on

previous attempts by geometers to create universal measuring devices, especially a measuring table developed by Leonhard Zubler in the 1620s.

35. Gorman, "The Angel and the Compass," 245–47; Baldwin, "Athanasius Kircher"; Hine, "Athanasius Kircher and Magnetism."
36. Gorman, "The Angel and the Compass," 250–51. Today magnetic declination is defined as the angle between "true north" and "magnetic north," but in this period, no one made this distinction.
37. To make matters more complicated, early modern observers often used the terms *declination* and *variation* interchangeably. To minimize confusion, I will simply use *declination*.
38. Fara, "Hidden Depths," esp. 572. For an introduction to "hollow earth" theories, including Kircher's, see Fitting, *Subterranean Worlds*.
39. Fara, "Hidden Depths," 572; Kollerstrom, "The Hollow World of Edmund Halley"; Warner, "Terrestrial Magnetism," esp. 74.
40. A. Cook, "Edmund Halley," 483–84. He used mainly the logs of ships for his cache of data and told Leibniz in 1703 that he was "still hoping to make a magnetic survey of the Pacific Ocean" as well.
41. Lorraine Daston notes that Halley's synoptic map and explanation of wind patterns stands as an emblem for the aspirations of those involved in collective programs of scientific observation in the seventeenth and eighteenth centuries. Daston, "The Empire of Observation," 91–92. See Halley, "An historical account."
42. Leibniz, "Concept einer Denkschrift," 247.
43. Ibid.
44. Leibniz, "Letter to Oldenburg," 169.
45. Gilbert, *On the Magnet*, 184.
46. Quoted in Bryden, "Magnetic Inclinatory Needles," 21. See also Bond, *The Longitude Found*.
47. Bryden, "Magnetic Inclinatory Needles," 22.
48. See Gordin, "Importation," 6; and Meyer, *Leibniz*, 42.
49. Leibniz, "Bedenken," 166.
50. Conrad Mel developed a "nautical pantometer" for use in conjunction with his own method for measuring longitude. He also invented a measuring device for use on land called a *Pantometrum Carolinum*, which he described as a newly invented machine for measuring the length and curves of paths, sketching their acclivities and declivities, and creating a proper profile of roads. Vial, *Dr. Conrad Mel*, 43. For a description of the nautical pantometer see Mel, *Pantometron Nauticum*. Christian Wolff and Leonhard C. Sturm criticized his efforts. For Sturm's evaluation for the Berlin Academy see BBAW/A I-V, 5a, Gutachten, and *Project de la Resolution* [1720].
51. Raspopov and Meshcheryakov, "Magnetic Declination," esp. 1147. Despite his best intentions, Leibniz had been unable to meet Peter I during the tsar's travels through Europe in 1697; however, in 1711 Leibniz attended the wedding of the tsar's son in Torgau, and the two became fast friends. They

met again in Carlsbad in 1712, after which the tsar made Leibniz a salaried counselor. Gordin, "Importation," 6.

52. Merrill, *Our Magnetic Earth*. Leibniz's own interest in magnetism, including his efforts to better understand the history and subterranean structure of the earth emerged as a result of the time he spent doing administrative work in mines in the Harz Mountains. See Hamm, "Knowledge from Underground"; Cohen and Wakefield, introduction to *Protogaea*, esp. xiv–xviii.

53. C. Eberhard, *Versuch*, 9.

54. Semler was also likely familiar with the writings of Athanasius Kircher. The Orphanage's archive possesses excerpts of his Mundus Subterraneus and more. See AFST/H E 51: 2–4.

55. After his first trip to Holland in 1697, Tsar Peter returned there several times; his final stay in Amsterdam was in 1717.

56. C. Eberhard, *Versuch*, 10. For more on 's Gravesande see Roberts, "Mapping Steam Engines," 197–218, esp. 199–201.

57. C. Eberhard, *Specimen Theoriae Magneticae*.

58. According to Eberhard's account, the instrument was "seen by Lord Pembroke and Desaguliers," who initially determined it was worthy of testing, as did Edmund Halley, William Whiston, Francis Hauksbee, and an anonymous clockmaker. C. Eberhard, *Versuch*, 11. John Theophilius Desaguliers joined the Royal Society in 1714 and served as an assistant to Isaac Newton. Francis Hauksbee was a curator of experiments and instrument maker for the Royal Society.

59. Dreyhaupt, *Pagus Neletici et Nudzici*, 719. Many helped Whiston gather more observations with the instrument at sea, Dreyhaupt continued, including the king, who gave him 5,000 thaler.

60. Whiston was a controversial figure largely because of his Arianism, or anti-Trinitarianism, which had resulted in his dismissal from a professorship at Cambridge in 1710. See Snobelen and Stewart, "Making Newton Easy"; Snobelen, "William Whiston"; Stewart, *The Rise of Public Science*; and Force, *William Whiston*.

61. Whiston, *The Longitude and Latitude Found*, xxv. Also mentioned in Fara, *Sympathetic Attractions*, appendix.

62. Whiston, *The Longitude and Latitude Found*, xxv. Eberhard seems to have been unaware of Henry Bond's book and ensuing discussion.

63. "He inform'd me also what Observations he had made by a larger Instrument of that Kind; how the Dip stood when he was young; and how much it had increas'd in his Time: for which and other the like Kindnesses ever since, I and the Publick are greatly obliged to him." In ibid., xxvi. See also Farrell, *William Whiston*, 148.

64. See Noël, *Observationes*. For his biography and experiences in China see Rule, "François Noël."

65. Whiston, *The Longitude and Latitude Found*, xxiv.

66. Ibid.

67. Ibid.
68. C. Eberhard, *Versuch*, 8–9; Noël's *Observationes* does not contain a hypothesis about a hollow earth. It consists of eight chapters that describe the eclipses of Jupiter's satellites, solar and lunar eclipses, latitudes and longitudes, the stars of the southern hemisphere, Chinese astronomy and units of measurement, magnetic inclination and declination, and comets and other miscellany.
69. C. Eberhard, *Versuch*, 13–14.
70. Ibid., 5.
71. Ibid., 7. Eberhard must have had the opportunity to discuss comets during his conversations with Halley and Whiston. Whiston argued for the agency of comets; he said they were the primary cause of the formation of the earth and the Great Flood described in the Bible. See Whiston, *A New Theory of the Earth*, and *Sir Isaac Newton's Mathematick Philosophy*; and Schechner, *Comets*, esp. 156–77 ("Halley's Comet Theory").
72. C. Eberhard, *Versuch*, 7.
73. Huygens, *Kosmotheoros*.
74. Gordin, "Importation," 23.
75. "Analogy is the instantiation of Reason in the Kosmotheoros: '. . . This must be our Method in this Treatise, wherein from the Nature and Circumstances of that Planet which we see before our eyes, we may guess at those that are farther distant from us.'" Ibid., 24.
76. C. Eberhard, *Versuch*, 9.
77. Ibid.
78. Ibid.
79. Like Edmund Halley, he was interested in making a case for his theory independent of scriptural authority, which made him vulnerable and likely to be accused of "atheism." See Schaffer, "Halley's Atheism."
80. Cohen and Wakefield, introduction to *Protogaea*, xix.
81. Ibid.
82. Ibid.
83. Ibid., xx–xxi.
84. Ibid., xxi. Even though Leibniz made use of the biblical account of the Great Flood in the *Protogaca*, he described it as one of many floods—not to mention earthquakes, volcanoes, and more—that contributed to the formation of the earth.
85. Rossi, *The Dark Abyss of Time*, esp. 33–56; Futch, "Leibniz on Plenitude," 541–42; Cohen and Wakefield, introduction to *Protogaea*, xxi. Burnet believed that since the flood, the earth had existed in a state of decline or deterioration; Leibniz, on the other hand, refused to accept this. Instead, he observed, there was a larger purpose, and a general movement toward perfection, observable in the earth's "ruins." Burnet also rejected the idea that the world was eternal; he insisted the world began when the earth was created. Leibniz, however, opened the door to accusations—most forcefully delivered by

Samuel Clarke—that he believed in an eternal or "beginning-less" world. Despite his efforts to defend his contention "that perfection can be maximized in a world that has a beginning but no end" (Futch, "Leibniz on Plenitude," 542), Leibniz's system of metaphysics, including his cosmogony, remained associated with this idea.

86. Schille, "Christoph Eberhard," 239.
87. Ibid.
88. Wotschke, "Das pietistische Halle," 475.
89. Wolff, "Gutachten," 239.
90. Wolff, *Allerhand Nützliche Versuche*, 113–232.
91. Ibid., 213–14.
92. Semler, *Methodus Inveniendae Longitudinis Maritimae*, 8.
93. Dreyhaupt, *Pagus Neletici et Nudzici*, 719.
94. Ibid.
95. See AFST/H A 173: 1, Tagebuch von A. H. Francke.
96. Winter, *Halle als Ausgangspunkt*, 65–66.
97. Semler, *Methodus Inveniendae Longitudinis Maritimae*, 9.
98. Ibid., 28.
99. Kramer, *August Hermann Francke*, pt. 2, 330.
100. Schille, "Christoph Eberhard," 239.
101. Ibid. In 1753, Johann Paul Eberhard published a description of a new "measuring table" (*Messtafel*) for measuring fields in any location; see J. Eberhard, *Beschreibung*.
102. *Wöchentliche Hallische Frage- und Anzeigungsnachrichten*, 231–34.
103. Bates and Jackson, *Glossary of Geology*, 329.
104. Howarth, "Fitting Geomagnetic Fields," 63–84.

CHAPTER 6

1. Francke, "Was noch aufs künftige projectiret ist," 501–2.
2. Gordin, "Importation," 6.
3. R. Wilson, "Replication Reconsidered," 203.
4. See Goldgar's discussion in *Impolite Learning*, esp. 19–21. I am thinking specifically of the work of Marcel Mauss and Claude Lévi-Strauss. See Mauss, *The Gift*; and Wiseman, *The Cambridge Companion to Lévi-Strauss*.
5. Goldgar, *Impolite Learning*.
6. The Halle Orphanage had its own newsletter, too, called the *Hallesche Berichte*, which mainly provided news about the mission to India. Rekha Kamth Rajan and Ulrike Gleixner have suggested that the *Halle Reports* were modeled after the Jesuits' *Lettres édifiantes et curieuses écrites*, a thirty-six-volume collection of letters from Jesuit missionaries first compiled and edited by Charles Le Gobien and published in Paris from 1703 to 1776. See Rajan, "Der Beitrag der Dänisch-Halleschen Missionare"; and Gleixner, "Expansive Frömmigkeit."

7. Bethmann, "Geschichte der Anstalt," 7–8.
8. Buchen, *Die gnädige Vorsorge Gottes*, 104–5.
9. Bethmann, "Geschichte der Anstalt," 10.
10. For example, *Nachricht von der gütigen Vorsorge Gottes*.
11. Buchen, *Die gnädige Vorsorge Gottes*, 39.
12. Ibid., 26–27.
13. Bethmann, "Geschichte der Anstalt," 14.
14. Buchen, *Die gnädige Vorsorge Gottes*, 111.
15. AFST/H A 170: 42, Christoph Andreas Chryselius to A. H. Francke, May 5, 1717.
16. Bethmann, "Geschichte der Anstalt," 11–14.
17. AFST/M 3 L 8: 140, Johann Wilhelm Kruckenberg to G. A. Francke, January 15, 1739.
18. Bethmann, "Geschichte der Anstalt," 16.
19. Carpzov, *Analecta*, 127–28.
20. Grünwald, *Ausführliche Beschreibung des Zittauischen Waysenhauses*, 30.
21. Ibid.
22. Ibid., 105–6.
23. Hinrichs, *Preußentum und Pietismus*, 340–42.
24. Ibid.
25. Ibid.
26. Ibid.
27. Ibid.
28. Steinbart, *Warhafftige und umständliche Nachricht*, 167.
29. Ibid., 156.
30. Ibid., 168.
31. *Leben des hochgelahrten Herr Johann Julius Hecker*, 49. For more on Hecker see Bloth, *Johann Julius Hecker (1707-1768)*; and Mentzel, *Pietismus und Schule*, esp. 110–64, and "Ein erfolgreicher Pietist an König Friedrichs Hof?"
32. See Hecker, *Betrachtung*, and *Einleitung in die Botanic*.
33. *Leben des hochgelahrten Herr Johann Julius Hecker*, 68–69.
34. Ibid., 73.
35. Ibid., 77–78.
36. Mentzel, *Pietismus und Schule*, 124.
37. Ibid., 73–74.
38. Ibid., 74.
39. Ibid., 131.
40. Ibid.
41. In ibid., 132; see also Nicolai, *Über meine gelehrte Bildung*.
42. Zippel, *Geschichte*, 7–8.
43. Ibid., 9.
44. Ibid., 10.
45. AFST/H C 16: 11, Theodor Gehr to A. H. Francke, February 14, 1698.
46. AFST/H C 16: 14–15, Theodor Gehr to A. H. Francke, May 2 and June 16, 1698.

47. Zippel, *Geschichte*, 13.
48. Ibid., 22–35.
49. Ibid., 44.
50. Ibid, 63.
51. Ibid., 66.
52. See Zippel, *Geschichte*, 74–83; and Stab/F 18, 1/2: 9, Vorschläge.
53. See, for example, AFST/M 1 C 16: 49, A. H. Francke to B. Schultze et al., September 19, 1726.
54. Fehr, *"Ein wunderlicher Nexus Rerum,"* 17.
55. Ibid., 18–19.
56. Klemme, *Die Schule Immanuel Kants*, 65; *Nachricht von den jetzigen Anstalten des Collegii Fridericiani.*
57. Fehr, *"Ein wunderlicher Nexus Rerum,"* 48.
58. Zippel, *Geschichte*, 88.
59. Ibid., 111.
60. Ibid., 104.
61. Voltaire, *Philosophical Dictionary*, 1:166–70.
62. Schrader, *Geschichte*, esp. 1:211–54.
63. Kleinert, "Johann Joachim Lange," 478.
64. Ibid., 478–79. See J. Lange, *Elementa philosophiae naturalis dogmaticae et experimentalis*; and Linne, *Systema Naturae*. See also Müller-Bahlke, *Wunderkammer*, 302.
65. Kleinert, "Johann Joachim Lange," 480.
66. Quoted in Winter, *Halle als Ausgangspunkt*, 343.
67. Semler continued to run a small workshop for making globes out of his own home. See S. Lange, *Leben Georg Friedrich Meiers*, 20–21; and AFST/H A 173: 1 Tagebuch von A. H. Francke.
68. See Whitmer, "What's in a Name?"; and te Heesen, "Boxes in Nature."
69. See Grote, "Pietistische Aisthesis," and "Moral Philosophy."
70. Zahn, *Erste Nachricht*, 13–14. For more on this institution see Stolzenburg, *Geschichte des Bunzlauer Waisenhaus.*
71. Zahn, *Erste Nachricht*, 13–14.
72. Sheehan, *The Enlightenment Bible*, 58–59.
73. Ibid., 59.
74. Schmidt, "On the Importance of Halle," 93.
75. Ibid., 94. See also similar remarks by Gawthrop in "Pedagogy and Reform."
76. See Basedow, *Philalethie.*
77. Te Heesen, *The World in a Box*, 73–74.
78. Kant, "Letters on the Philanthropinum at Dessau," 245.
79. For a recent study of the *Philanthropinum*, Basedow, and other late-eighteenth-century pedagogues' preoccupation with teaching children how to see through the systematic observation of materials and scientific instruments, see Ramirez Jasso, "Imagining the Garden."

Bibliography

Archival Sources

ARCHIVE DER FRANCKESCHEN STIFTUNGEN

/ B: *Bibliothekarchiv*
/ H: *Hauptarchiv*
/ M: *Missionsarchiv*
/ W: *Wirtschaftsarchiv*

AFST/ B 50: 00. Einige Betrachtungen von der von der möglichen Vereinbahrung der Vernünft und Religion, vorgeleget von Boyle (a few observations about the reconcileableness of reason and religion, presented by Boyle).

AFST/ H A 170: 42. Letter from Christoph Andreas Chryselius to A. H. Francke [May 5, 1717].

AFST/ H A 173: 1. A. H. Francke's Diary [January 1–December 31, 1719].

AFST/ H A 183: 9. Letter from Peter Müller to A. H. Francke [April 5, 1710].

AFST/ H C 16: 11, 14–15. Letters from Theodor Gehr to A. H. Francke [February 14, May 2, and June 16, 1698].

AFST/ H C 147: a. Letter from Gottfried Wilhelm Leibniz to A. H. Francke [August 7, 1697]

AFST/ H C 147: f. Letter from Gottfried Wilhelm Leibniz to A. H. Francke [January 17, 1714].

AFST/ H C 229: 58. Letter from Anton Wilhelm Böhme to Carl Hildebrand von Canstein [August 29, 1710].

AFST/ H C 263: 1. Letter from E. W. von Tschirnhaus to A. H. Francke [1698].

AFST/ H C 265: 3 and 7. Letters from Christoph Eberhard to A. H. Francke [March 20, 1706, and April 25, 1710].

AFST/ H C 491. Briefe, Berichte, Schulordnung und Projektentwurf zur Gründung eines Waisenhauses in Tobolsk (Letters, Reports, School Regulations and Project Plans for the Founding of an Orphanage in Tobolsk).

AFST/H D 85, 599–626. Aufsatz von Ehrenfried Walther von Tschirnhaus über das Studium der Mathematik und Mechanik [1698] (Essay by E. W. von Tschirnhaus about the Study of Mathematics and Mechanics).

AFST/H D 99. Journal von Heinrich Julius Elers über seine Reise nach Dresden [1704] (Journal of Heinrich Julius Elers about his Trip to Dresden).

AFST/H E 51: 2. Athanasius Kircher, Mundus Subterraneus, Excerpta.

AFST/H E 51: 3. Athanasius Kircher, Tractatus de arte magna lucis et umbrae, Excerpta.

AFST/H E 51: 4. Athanasius Kircher, Epilogus sive Metaphysica lucis et umbrae tractatui de arte magna lucis et umbrae adjectus.

AFST/H E 61. Manuskript für den Realienunterricht mit den zu behandelten Inhalten von Christoph Semler (Object lessons manuscript, with contents by C. Semler).

AFST/H F 14: 289. Letter from Michael Bel[c]k to A. H. Francke [June 14, 1708].

AFST/H L 16: 4 and 5. Früh-Predigt von August Hermann Francke über das Johannesevangelium, Kapitel 1, gehalten Fes. 3 Nativit. "Philanthropia Dei" [December 27, 1704] (Early Sermon by A. H. Francke about . . . "Philanthropia Dei").

AFST/M 1 C 1: 27. Letter from Bartholomäus Ziegenbalg to A. H. Francke [October 1, 1706].

AFST/M 1 C 1: 35. Letter from Bartholomäus Ziegenbalg and Heinrich Plütschau to Joachim Justus Breithaupt, Paul Anton, and A. H. Francke [October 17, 1707].

AFST/M 1 C 16: 49. Letter from August Hermann Francke to Benjamin Schultze, [Nikolaus Dal, Christian Friedrich Pressier, Christoph Theodosius Walther], and [Martin Bosse] [September 19, 1726].

AFST/M 2 A 5. Letters exchanged between August Hermann Francke, Gotthilf August Francke, and Maturin Veyssière La Croze and additional materials.

AFST/M 3 L 8: 140. Letter from Gotthilf August Francke to Johann Wilhelm Kruckenberg [October 1739].

AFST/W VII/I/20. Instruktionen für die Herumführer [1732, 1746] (Instructions for Tour Guides).

AFST/W XI/-/71. Inventarverzeichnisse der Kunst- und Naturalienkammer (Catalog of the Chamber of Artificial and Natural Things).

AFST/W XI/-/72. Demonstration des Modells der Stiftshütte Mosis (Demonstration of the Model of the Tabernacle).

STAATSBIBLIOTHEK (PREUßISCHER KULTURBESITZ) / NACHLASS A. H. FRANCKE [BERLIN]

Stab/F 4b/13: 40. Letter from Maria Sophie von Marschall to A. H. Francke [January 25, 1699].

Stab/F 7/14: 37 and 38. Letters from Johann Franz Buddeus to A. H. Francke [February 2 and 16, 1719].

Stab/F 15, 1/5: 4, 10. Letters from Conrad Mel to A. H. Francke [June 15, 1709, and January 2, 1711].

Stab/F 18, 1/2: 9. Vorschläge, wie das Collegium Fridericanum in Königsberg mit den gleichen Privilegien wie die Glauchaschen Anstalten ausgestattet werden kann (Suggestions regarding how the Collegium Fridericanum in Königsberg can be given the same privileges as the Glaucha institutions).

Stab/F 22/5. Letter from Gottfried Vockerodt to A. H. Francke [December 17, 1710].

Stab/F 27/13: 2. Letter from Johann Hosmann von Rothenfels to A. H. Francke [April 12, 1708].

Stab/F 28/1: 2–4. Ergänzende Fragen Georg Heinrich Neubauers zum Studium der Waisenhäuser in Holland [1697] und Aufzeichnungen (Supplementary Questions of Georg Heinrich Neubauer about the Study of Orphanages in Holland and Drawings).

Stab/F 30/47: 1–10. Letters from Frederick Slare to A. H. Francke [ca. 1704–22].

Stab/F 33/2: 2. Letter from Heinrich Plütschau to A. H. Francke [October 16, 1706].

GEHEIMES STAATSARCHIV PREUßISCHER KULTURBESITZ [BERLIN]

GSPK, I HA Rep 52 Nr 159b. Letters: Friedrich Wilhelm I, 1721–49.

BERLIN-BRANDENBURGISCHE AKADEMIE DER WISSENSCHAFTEN/ARCHIV [BERLIN]

BBAW/A I–V, 5a. Gutachten (Evaluation) by Leonhard Christoph Sturm, 1708.

Primary Sources (Printed)

Arndt, Johann. *Wahres Christentum Vier Bücher* [1605–1610]. Reprinted as *True Christianity,* edited and translated by Peter Erb. Mahwah, NJ: Paulist Press, 1979.

Arnold, Gottfried. *Vitae Patrum Oder das Leben der Altväter.* Halle: Waysenhaus, 1700.

———. *Göttliche Liebes-Funcken,* parts 1 and 2. Frankfurt: Zunner, 1701.

Bahr, Hieronymous. *Höchstverderbliche Auferziehung der Kinder bey den Pietisten, durch Gelegenheit des von dem hällische Professore M. August Herrmann Francken canonisirten zehen-jährigen Kindes Christlieb Leberecht Exters.* S.l., 1709.

Basedow, Johann. *Philalethie. Neue Aussichten in die Wahrheiten und Religion der Vernunft.* Altona: Iversen, 1764.

Bayly, Lewis. *The Practice of Piety, directing a Christian how to walk that he may please God.* London, 1613 (date of first printing unknown). Reprinted and translated as *Praxis pietatis: Das ist Übung der Gottseligkeit.* Lüneburg: Stern, 1634.

Bessler, Johann Ernst. *Der Rechtglaubige Orffyreer: Oder die Einige Vereinigung der uneinigen Christen in Glaubens Sachen. Sie nennen sich gleich; Evangelisch-Lutherisch, Evangelisch-Reformirt, Römisch Catholisch oder Papistisch.* Kassel: Cramer, 1723.

Boecler, Johann Heinrich. *Characteres Politici In Velleio Paterculo expositione quadam demonstrati.* Strasbourg: Staedel, 1672.

Böhme, Jakob. *Das Sechste Büchlein: Vom Ubersinnlichen Leben, Nebst ein Gespräche eines Meisters und Jüngers: Wie die Seele möge zu Göttlicher Anschauung und Gehör kommen . . . .* In *Der Weg zu Christo: Verfasset in neun Büchlein.* Amsterdam, 1682.

Bond, Henry. *The Longitude Found, or a treatise showing an easy and speedy way, as well as by night as by day, to find the Longitude, having but the Latitude of the place and the Inclination of the Inclinatory Needle.* London: Bond, 1676.

Boyle, Robert. *Seraphic Love, or Some Motives and Incentives to the Love of God* [1659/1660]. Reprinted as "Seraphic Love" in *The Works of Robert Boyle*, vol. 1, edited by Michael Hunter and Edward B. Davis, 51–134. London: Pickering and Chatto, 1999.

———. *Some Considerations Touching the Style of the Holy Scriptures* [1661]. Reprinted as "Some Considerations Touching the Style of the Holy Scriptures," in *The Works of Robert Boyle*, vol. 2, edited by Michael Hunter and Edward B. Davis, 379–490. London: Pickering and Chatto, 1999.

———. *Excellence of Theology compared with Natural Philosophy* [1664]. Reprinted as "Excellence of Theology Compared with Natural Philosophy," in *The Works of Robert Boyle*, vol. 8, edited by Michael Hunter and Edward B. Davis, 3–98. London: Pickering and Chatto, 2000.

———. *Some Considerations about the Reconcileableness of Reason and Religion* [1675]. Reprinted as "Some Considerations about the Reconcileableness of Reason and Religion," in *The Works of Robert Boyle*, vol. 8, edited by Michael Hunter and Edward B. Davis, 233–94. London: Pickering and Chatto, 2000.

———. *Treatises on the High Veneration Man's Intellect Owes to God* [1684–1685]. Reprinted as "Treatises on the High Veneration Man's Intellect Owes to God," in *The Works of Robert Boyle*, vol. 10, edited by Michael Hunter and Edward B. Davis, 157–204. London: Pickering and Chatto, 2000.

———. *Amor Seraphicus. Die Seraphische Liebe, oder einige Anreissungen zur Liebe gegen Gott.* Halle: Waysenhaus, 1708.

———. *Auserlesene Theologische Schrifften Des Edlen Roberti Boyle, Weyland Mitglieds der Königl. Societaet der Wissenschaften in England Auserlesene Theologische Schrifften: (1.) Dessen Gedancken vom Stilo und Schreib-Art der heil. Schrift; (2.) Von der Vortrefflichkeit der Theologie in Vergleichung mit der Philosophie; (3.) Von der Veneration und Verehrung so der menschl. Verstand Gott schuldig ist; (4.) Von der Seraphischen Liebe; Nunmehro Wegen ihrer Würdigkeit zum gemeinen Nutzen ins Teutsche übersetzet und mit gehörigem Register versehen.* Halle: Waysenhaus, 1709.

Buchen, Christoph. *Die gnädige Vorsorge Gottes, in einer wahrhafftigen Nachricht von dem Waysen-Hause bey Weissenfels an Langendorff gelegen: Welches ein armer*

Fuhrmann so wohl vor Knaben als Mädgen erbauet; Daß dieselben darinne im Christenthum und andern nützlichen Wissenschaften unterrichtet, und zu allerhand Arbeit angewöhnet werden. Leipzig, 1721–23.

Buddeus, Johann Franz. *Elementa philosophiae practicae.* Halle: Zeidler 1697; 3rd edition, 1707. Reprinted in *Gesammelte Schriften, Johann Franz Budde,* vol. 3, edited by Walter Sparn. Hildesheim: Olms-Weidmann, 2004.

———. *Introductio ad historiam philosophiae Ebraeorum.* Halle: Waysenhaus, 1702. Reprinted in *Gesammelten Schriften Johann Franz Budde,* vol. 4, edited by Walter Sparn. Hildesheim: Olms-Weidmann, 2004.

———. *Elementa philosophiae instrumentalis seu institutionum philosophiae eclecticae tomus primus.* Halle: Waysenhaus, 1703.

———. *Elementa philosophiae theoreticae, seu institutionum philosophiae eclecticae tomus secundus.* Halle: Waysenhaus, 1703.

———. *Bedencken über die Wolffianische Philosophie.* Frankfurt am Main: Andreä, 1724.

Carpzov, Johann Benedict. *Analecta Fastorum Zittauiensium der Historischer Schauplatz der löblichen Alten Sechs Stadt des Margraffthums ober Lausitz. . . .* Leipzig, 1716.

Comenius, Johann Amos. *Naturall philosophie reformed by divine light: or, a synopsis of physicks [Physicae ad lumen divinum reformatae synopsis (1633)].* London: Leybourn, 1651.

———. *De Rerum Humanarum Emendatione Consultatio Catholica, Ad Genus Humanum Ante alios vero Ad Eruditos, Religiosos, Potentes Europae.* Halle: Waysenhaus, 1702.

———. *Historia Fratrum Bohemorum, Eorum Ordo Et Disciplina Ecclisiatica, ...* With a preface by Johann Franz Buddeus. Halle: Waysenhaus, 1702.

Dilherr, Johann Michael. *Propheten Schul. Das ist, Christliche Anweisung zu Gottseliger Betrachtung des Lebens und der Lehre Heiliger Propheten Alten Testaments.* Nuremberg: Fürst, 1662.

Dreyhaupt, Johann Christoph. *Pagus Neletici et Nudzici, oder Ausführliche diplomatisch-historische Beschreibung des zumehemaligen Primat und Erz-Stifft nunmehr aber durch den westphälischen Friedens-Schluss secularisirten Herzogthum Magdeburg gehörigen Saal-Kreyses. . . .* Halle: Schneider, 1750.

Eberhard, Christoph. *Specimen Theoriae Magneticae. Quo ex certis Principiis Magneticis ostendibus vera et universalis Methodus inveniendi Longitudinem et Latitudinem.* London, 1718.

——— [Alethophilus, pseud.]. *Der innere und äussere Zustand derer Schwedischen Gefangenen in Rußland, durch Ihre Brieffe, darinnen Sie-ihre Bekehrung, wie die Aufrichtung der öffentl. Schule zu Tobolsky der Hauptstadt in Sibirien melden.* Frankfurt and Leipzig, 1718–21.

———. *Versuch einer Magnetischen Theorie, in welchem nach gewissen Grund-Sätzen Anleitung gegeben wird, Den rechten und allgemeinen Weg zur Länge und Breiter der Oerter, so wol auf der See als zu Lande, vermittelst des Magnets zu finden.* Leipzig: Martini, 1720.

Eberhard, Johann Paul. *Beschreibung einer neuen Meßtafel vermittelst welcher man mit geringen Kosten aller Orten Feldmessen kann: nebst einen Anhang vom Gebrauch der Nepperischen Rechenstäbe*. Halle: Hemmerde, 1753.

Francke, August Hermann. "Historische Nachricht von Verpflegung der Armen und Erziehung der Jugend in gedachtem Glaucha." In *Christliche Verpflegung der Armen*, by Philip Jakob Spener, appendix. Frankfurt an der Oder: Schrey, 1697.

———. "Brief an P. J. Spener" [Glaucha, January 8, 1698]. In *Beiträge zur Geschichte A. H. Franckes: enthaltend den Briefwechsel Francke's und Spener's*, compiled and edited by Gustav Kramer, 380-81. Halle: Buchhandlung des Waysenhauses, 1861.

———. "Die Lehre von der Erleuchtung" [March 6, 1698]. In *August Hermann Francke Predigten*, vol. 1, edited by Erhard Peschke, 380-99. Berlin: Walter De Gruyter, 1987.

———. *Einrichtung des Pädagogii zu Glauch an Halle*. Halle: Waysenhaus, 1700.

———. "Projekt zu einem Seminario Universali oder Anlegung eines Pflanz-Gartens" [1701]. In *August Hermann Francke—Werke in Auswahl*, edited by Erhard Peschke, 108-15. Berlin: Evangelische, 1969.

———. "Kurzer und einfältiger Unterricht: wie die Kinder zur wahren Gottseligkeit und christlichen Klugheit anzuführen sind" [1702]. In *August Hermann Francke's Pädagogische Schriften*, compiled and edited by D. G. Kramer, 15-71. Langensalza: Beher and Söhne, 1885.

———. *Ordnung und Lehrart, wie selbige in dem Pädagogio zu Glaucha an Halle eingeführet ist, worinnen vornehmlich zu befinden, wie die Jugend nebst der Anweisung zum Christentum in Sprachen und Wissenschaften, als in der lateinischen, griechischen, ebraeischen und französischen Sprache wie auch in Calligraphia, Geographia, Historia, Arithmetica, Geometria, Oratoria, Theologia und in denen fundamentis Astronomicis, Botanicis, Anatomicis etc. auf eine kurze und leichte Methode zu unterrichten und zu denen Studiis Academicis zu praeparien sei*. Halle: Waysenhaus, 1702.

———. *Ordnung und Lehr-Art, wie selbige in denen zum Wäysen-Hause gehörigen Schulen eingeführet ist: Worinnen vornemlich zu befinden, Wie die Kinder in und ausser der Schul in Christlicher Zucht zu halten, und zum Lesen, zierlichen Schreiben, Rechnen, wie auch zur Music und andern nützlichen Dingen anzuführen sind*. Halle: Waysenhaus, 1702.

———. "Projekt des Collegii orientalis theologici abgefasset in Majo 1702." In *August Hermann Francke: Ein Lebensbild*, part 1, compiled and edited by Gustav Kramer, 278-85. 1880. Reprint, Hildesheim: Georg Olms, 2004.

———. *Schriftmäßige Betrachtung von Gnade und Wahrheit: . . . Dem in dieser neuen Edition noch beygefüget worden die vorhin besonders gedruckte Betrachtung von der Philanthropia Dei oder Liebe Gottes gegen die Menschen*. Halle: Waysenhaus, 1705. Reprinted in 1721, 1723, 1727, and 1729 and as "Philanthropia Dei, das ist die Liebe Gottes gegen die Menschen," in *August Hermann Francke's Pädagogische Schriften*, compiled and edited by D. G. Kramer, 93-97. Langensalza: Beyer, 1885.

———. *Philotheïa: Oder die Liebe zu Gott.* Halle: Waysenhaus, 1706, 1723. Reprinted as "Philotheïa oder die Liebe zu Gott," in *August Hermann Francke's Pädagogische Schriften,* compiled and edited by D. G. Kramer, 98–106. Langesalza: Beyer, 1885.

———. *La Philanthropie ou L'Amour de Dieu envers les Hommes: Propose aus jeunes gens, qui êtudient dans les Ecoles dependantes de la Maison des Orfelins, à la fin d'un examen, pour le mediter soigneusement, & s´exciter à la pieté par Auguste Armand Francke, Professeur en Theologie à Halle, . . . Traduit de l'Allemand.* Halle: Maison des Orfelins, 1709.

———. "Was noch aufs künftige projectiret ist . . ." [1711]. Reprinted in *August Hermann Francke: Ein Lebensbild,* part 2, by Gustav Kramer, 498–503. Halle: Buchhandlung des Waisenhauses, 1882. Reprint, Hildesheim: Georg Olms, 2004.

———. "Briefe an Gottfried Wilhelm Leibniz" [Halle, January 10 and February 17, 1714]. Printed in "Die Russlandthematik im Briefwechsel zwischen August Hermann Francke und Gottfried Wilhelm Leibniz," by Gerda Utermöhlen, in *Hallesche Forschung,* vol. 1, *Halle und Osteuropa zur europäische Ausstrahlung des hallischen Pietismus,* edited by Johannes Wallmann and Udo Sträter, 124–28. Tübingen: Max Niemyer, 1998.

———. *Die Klugheit der Kinder des Lichts.* Halle: Waysenhaus, 1714. Reprinted in *August Hermann Francke Predigten und Tractätlein,* part3. Halle: Waysenhaus, 1723. Reprinted in *August Hermann Francke Sonn und Fest-Tags Predigten.* Halle: Waysenhaus, 1724.

———. *Das Auge des Glaubens.* Halle: Waysenhaus, 1716.

———. "Brief an Friedrich II" [August 1720]. In *August Hermann Francke: Ein Lebensbild,* part 1, compiled and edited by Gustav Kramer [First edition, 1880], 326–27. Hildesheim: Georg Olms, 2004.

———. *Lectiones paraeneticae oder öffentliche Ansprachen an die studiosos theologiae auf der Universität zu Halle in dem so genannten collegio paraenetico.* Halle: Waisenhaus, 1726–29.

Freyer, Hieronymous. *Verbesserte Methode des Pädagogii Regii zu Glaucha vor Halle.* Halle: Waysenhaus, 1721. Reprinted as "Verbesserte Methode des Paedagogii Regii zu Glaucha vor Halle," in *August Hermann Francke's Pädagogische Schriften,* compiled and edited by D. G. Kramer, 357–436. Langesalza: Beyer, 1876.

Gilbert, William. *On the Magnet and Magnetic Bodies, and on that Great Magnet the Earth* [1600]. Edited by Derek J. Price, translated by Silvanus Phillips Thompson. New York: Basic Books, 1958.

Gründlicher Unterricht von der Graphice. Halle: Waysenhaus, 1717.

Grünwald, M. Martin. *Ausführliche Beschreibung des Zittauischen Waysenhauses, und dessen Loblichster Einrichtung aus heiligen Absehen entworffen und allen welchen Gottes Väterlichen Vorsorge gebührend verwundern zu danckbarhere Uberlegung.* Zittau: Richter, 1710.

Hall, Joseph. *Characteres der Menschen oder die Entlarvete Welt.* Amsterdam: Siebert Siebertsen, 1701.

Hallesche Berichte (Halle reports), das heisst Der Königl. *Dänischen Missionarien aus Ost-Indien eingesandter Ausführlichen Berichten, Von dem Werck ihres Amts unter den Heyden*. . . . Parts 1–9 (Continuation 1–108). Halle: Waisenhaus, 1710–72.

Halley, Edmund. "An historical account of the trade winds and monsoons observable in the seas between and near the tropicks, with an attempt to assign the cause of the said winds." *Philosophical Transactions of the Royal Society of London* 16 (1686): 153–68.

———. "An account of the cause of the change of the variation of the magnetic needle; with an hypothesis of the structure of the internal parts of the earth." In *Philosophical Transactions of the Royal Society of London* 16 (1692): 563–78.

Hecker, Johann Julius. *Betrachtung des menschlichen Cörpers nach der Anatomie und Physiologie*. Halle: Waysenhaus, 1734.

———. *Einleitung in die Botanic*. Halle: Waysenhaus, 1734.

Herrnschmidt, Johann Daniel. "Von den rechten Grenzen der natürlichen Philosophie" [1718]. In *Kurtze Fragen von denen natürlichen Dingen welche Gott als Zeugen seiner Liebe den Menschen vor Augen gestellet*, by Johann Georg Hoffmann, 5–66. Halle: Waysenhaus, 1720.

Hoffmann, Friedrich. "Vorrede." In *Einleitung in die Botanic,Worinnen die nöthigste Stücke dieser Wissenschaft kürtzlich abgehandelt werden*, by Johann Julius Hecker. Halle: Waysenhaus, 1734.

Hugo, Hermann. *Pia Desideria, elegantissimo carmine descripta: una cum Barlaei, aliorumque praestantiorum nostrae & superioris aetatis poëtarum sacris carminibus selectioribus, & brevi singulorum vitae historia*. Gotha, 1707.

Huygens, Christian. *Kosmotheoros: Celestial Worlds Discover'd; or Conjectures Concerning the Inhabitants, Plants and Productions of Other Worlds in the Planets*. London, 1698.

Kant, Immanuel. "Letters on the Philanthropinum at Dessau" [1777]. In *The Educational Theory of Immanuel Kant*, translated and edited by Edward Franklin Buchner, 242–46. Philadelphia and London: B. Lippincott, 1908.

Lange, Johann Joachim. *Elementa philosophiae naturalis dogmaticae et experimentalis*. Halle: Waysenhaus, 1735.

Lange, Samuel Gotthold. *Leben Georg Friedrich Meiers*. Halle: Gebauer, 1778.

Leben des hochgelahrten Herr Johann Julius Hecker. In *Ehrengedächniss des weiland Hochwürdigen und Hochgelahrten Herrn Johann Julius Hecker . . . Pastoris bey der Dreyfaltigkeitskirche und Directoris der Königl. Realschule und des Frankfurtschen Waisenhauses*. Berlin, 1769.

Leibniz, Gottfried Wilhelm. "Letter to Oldenburg" [September 18, 1670]. In *The Correspondence of Henry Oldenburg*, vol. 7, *1670-1671*, edited by A. Rupert Hall and Marie Boas Hall, 169–70. Madison: University of Wisconsin Press, 1970.

———. "Grundriss eines Bedenkens von Aufrichtung einer Societät" [1671]. In *Sämtliche Schriften und Briefe Gottfried Wilhelm Leibniz*, no. 4,*Politische Schriften*, vol. 1, *1667-1676*, edited by Paul Ritter, 530–43. Berlin: Akademie der Wissenschaften, 1983.

———. "Brief an A. H. Francke" [Hannover, August 17, 1697]. In *G. W. von Leibniz und die China-Mission*, by Franz Rudolf Merkel, 215–16. Leipzig: Hinrichs, 1920. Reprinted in *Sämtliche Schriften und Briefen Gottfried Wilhelm Leibniz*, no. 1, *Allgemeiner, Politischer und Historischer Briefwechsel*, vol. 14, *Mai bis Dezember 1697*, edited by Gerda Utermöhlen, Sabine Sellschopp, and Wolfgang Bungies, 397–400, letter no. 241. Berlin: Akademie der Wissenschaften, 1993.

———. "Brief an Spener" [Berlin, June 8, 1700]. In *Sämtliche Schriften und Briefen Gottfried Wilhelm Leibniz*, no. 1, *Allgemeiner, Politscher und Historischer Briefwechsel*, vol. 18, *January–August 1700*, edited by Malte-Ludolf Babin, 703–4, no. 398. Berlin: Akademie der Wissenschaften, 2004.

———. "Brief an Tschirnhaus" [Hannover, April 17, 1701]. In *Der Briefwechsel von Gottfried Wilhelm Leibniz mit Mathematikern. Tschirnhaus, Huygens, Newton*, edited by Carl Immanuel Gerhardt, 510–15. Berlin: Georg Olms, 1899. Reprint, Hildesheim: Olms, 1987.

———. "Bedenken: wie bey der Neuen Konigl. Societät der Wißenschafften, der allergnadigsten instruction gemäß, propagation fidei per scientias forderlichst zu veranstalten" [Berlin, November 1701]. In *Leibniz und seine Akademie: ausgewählte Quelle zur Geschichte der Berliner Sozietät der Wissenschaften, 1697–1716*, edited by Stephen Brather, 161–67. Berlin: Akademie, 1993.

———. "Concept einer Denkschrift Leibniz's über Unterssuchung der Sprachen und Beobachtung der Variation des Magnets im Russischen Reiche, No. 158" [1712]. In *Leibniz in seinen Beziehungen zu Russland und Peter dem Grossen*, by W. Guerrier, 239–49. St. Petersburg: Eggers, 1873.

———. "Brief an August Hermann Francke" [Vienna, January 17, 1714]. Printed in "Die Russlandthematik im Briefwechsel zwischen August Hermann Francke und Gottfried Wilhelm Leibniz," by Gerda Utermöhlen, in *Hallesche Forschung*, vol. 1, *Halle und Osteuropa zur europäische Ausstrahlung des hallischen Pietismus*, edited by Johannes Wallmann and Udo Sträter, 127–28. Tübingen: Max Niemyer, 1998.

———. *Protogaea* [1749]. Translated and edited by Claudine Cohen and Andre Wakefield. Chicago: University of Chicago Press, 2008.

Leibniz, Gottfried Wilhelm, and Gerhard Molanus. "De Unione Protestantium Molani et Leibnitii Judicium." In *Opera Omnia*, vol. 1, *Theologia*, edited by Louis Dutens, 735–37. Geneva, 1768. Reprint, Hildesheim: Georg Olms, 1989.

Linne, Carl von. *Systema Naturae, Sive Regna Tria Naturae Systematice Proposita Per Classes, Ordines, Genera Et Species. . . .* Halle: Gebauer, 1740.

Lund, Johannes. *Die Alten Jüdischen Heiligthümer, Gottesdienste and Gewohnheiten.* Hamburg: Liebernickel, 1701.

Mel, Conrad. "Die Schauburg der Evangelischen Gesandtschaft oder ohnmassgebliche Vorschläge wegen Fortpflanzung des allerhieligsten Glaubens durch Bekehrung der Heiden sonderlich in China . . ." [1701]. In *G. W. von Leibniz und die China-Mission*, by Franz Rudolph Merkel, 225–39. Leipzig: Hinrichs, 1920. Reprinted in *August Hermann Francke: Ein Lebensbild*, part 2 [1880], by Gustav Kramer, 285–303. Hildesheim: Georg Olms, 2004.

———. *Pantometron Nauticum, seu machina: Pro Invenienda Longitudine et Latitudine Locorum in Mari.* Hersfeld, 1707.

———. *Wäysen-Predigt: Vorstellend, das treue Vatter-Herz Gottes gegen arme verlassene Wäysen.* Hersfeld: Vogel, 1709.

———. *Missionarius Evangelicus: Seu Consilia, De Conversione Ethnicorum, Maxime Sinensium.* Leipzig: Cramer, 1711.

Nachricht von den jetzigen Anstalten des Collegii Fridericiani. Königsberg, 1741.

Nachricht von der gütigen Vorsorge Gottes oder kurtzer und aufrichtiger Bericht von der Auferbauung, bißheriger Unterhaltung und Zustand des bey Weißenfels zu Langendorf gelegenen Waysen-Hauses, mitgetheilt von einem Freund, der die Wahrheit bekennt. Jena, 1714. Reprint, Leipzig, 1716.

Neubauer, Georg Heinrich. *Was bey Erbauung unseres Waysen-Hauses zu wissen nöthig sey: der Fragenkatalog Georg Heinrich Neubauers für die Hollandreise 1697. Kleine Texte der Franckeschen Stiftungen Nr. 9.* Halle: Frankeschen Stiftungen, 2003.

Nicolai, Christoph Friedrich. *Über meine gelehrte Bildung.* Berlin, 1799.

Noël, François. *Observationes mathematicae et physicae in India et China factae . . . ab anno 1684.* Prague: Kamenicky, 1710.

Penther, Johann Friedrich. *Ausführlichen Anleitung zur bürgerlichen Baukunst,* part 2. Augsburg: Pfeffel, 1745.

Reyher, Andreas. *Kurtzer Unterricht: I. Von Natürlichen Dingen. II. Von etlichen nützlichen Wissenschafften. III. Von Geist- und Weltlichen Land-Sachen. IV. Von etlichen Hauß-Regeln; Auff gnädige Fürstl. . . .* Gotha, 1659, 1664, 1691.

Sabunde, Raymond de. *(Oculus Fidei) Theologia Naturalis; sive Liber Creaturarum.* Edited by Johann Amos Comenius. Amsterdam: Pieter van den Berge, 1661.

Scheuchzer, Johann Jacob. *Kupfer-Bibel, in welcher Die Physica Sacra, Oder Geheiligte Natur-Wissenschafft Derer in Heil. Schrifft vorkommenden Natürlichen Sachen.* Augsburg: Pfeffel, 1731.

Semler, Christoph. *Nützliche Vorschläge von Auffrichtung einer Mathematischen Handwercks-Schule bey der Stadt Halle. . . .* Halle: Henckel, 1708.

———. *Neueröffenete Mathematische und Mechanische Realschule. . . .* Halle: Renger, 1709.

———. *Der Tempel Salomonis: . . . nebst allen und jeden in folgender Beschreibung und beygefügten Kupferstücken enthaltenen Theilen desselben, in einem eigentlichen Modell und materiellen Fürstellung, in dem Wäysen-Hause zu Glaucha an Halle, zu Erläuterung sehr vieler Örter der Heiligen Schrift.* Halle: Waysenhaus, 1718.

———. *Die Stadt Jerusalem Mit allen ihren Mauren, Thoren, Thürmen, Tempel, Pallästen, Schlössern, auch übrigen publiquen und privat-Gebäuden, samt denen Thälern, Bergen und umliegenden Bergen, In einem Modell und materiellen Fürstellung aufgerichtet.* Halle: Waysenhaus, 1718.

———. *Palaestina Oder das Gelobte Land und dessen Berühmteste Städte, Wälder, Wüsten, Meere, Flüsse, Berge, Thäler, Hölen, Grabmahle, Gärten, Äcker, Weinberge, auch Gegenden derer umliegenden Länder in einem eigendliche Modell fürgestellet.* Halle: Waysenhaus, 1722.

———. *Eigentliches Modell und materielle Figur zu Erläuterung sehr vieler Örter Heiliger schrift aufgerichtet.* Halle: Waysenhaus, 1723.

———. *Supremo Magnae Britanniae Senaturi Illustrissimo Parlamento Consecrata Humillimeque Submissa Methodus Inveniendae Longitudinis Maritimaee per Acus Verticales Magneticas, Evidentissimis Superstructa Experimentis Terra Marique Factis.* Halae Magdeburgicae: Orphanotropheum, 1723.

Specification, derer Sachen, welche zu der für die Glauchische Anstalten angefangenen Naturalien-Cammer bis anhero verehret worden. 1700.

Spener, Philipp Jakob. "Pia Desideria: Oder Hertzliches Verlangen nach Gottgefälliger Besserung der wahren Evangelischen Kirchen" [Frankfurt, 1676]. Translated by Theodore G. Tappert. In *Seventeenth Century German Prose: Grimmelshausen, Leibniz, Opitz, Weise and Others,* edited by Lynne Tatlock, 93–104. New York: Continuum, 1993.

———. "Brief an Johann Jacob Spener in Leipzig" [Dresden, March 13, 1688]. In *Philipp Jakob Spener Briefe aus der Dresdner Zeit: 1686–1691,* vol. 2, edited by Johannes Wallmann, 133–36. Tübingen: J. C. B. Mohr Siebeck, 2008.

———. "Brief an Ehrenfried Walther von Tschirnhaus in Kieslingswalde" [September 20, 1688]. In *Philipp Jakob Spener Briefe aus der Dresdner Zeit: 1686–1691,* vol. 2, edited by Johannes Wallmann, 399–403. Tübingen: J. C. B. Mohr Siebeck, 2008.

———. *Theologische Bedencken und andere brieffliche Antworten auff geistliche, sonderlich zur Erbauung gerichtete Materien.* Halle: Waisenhaus, 1707–9.

Steinbart, Siegmund. *Warhafftige und umständliche Nachricht Derjenigen Tropffen, Strömlein und Flüsse, so aus Gottes reicher Seegens-Quelle in das von ihm selbst Vor der Stadt Züllichow bey Krausche Nicht so wohl Zu blosser Erzieh- und Unterhaltung armer verlassener Kinder.* Berlin: Schlechtigern, 1723.

Sturm, Johann Christoph. *De Philosophia sectarian et elective.* Nuremberg, 1679.

Sturm, Leonhard Christoph. *De sciagraphia templi Hierosolymitani.* Leipzig, 1694.

———. *Nicholas Goldmanns vollständige Anweisung zu der Civilbaukunst.* Leipzig, 1696.

———. *Kurtze Vorstellung der gantzen Civil Baukunst.* Augsburg: Wolf, 1718.

———. *Project de la Resolution du fameux Probleme touchant la longitude sur Mer.* Nuremberg: Monath, 1720.

"Temple." In *Grosses vollständiges Universal-Lexicon aller Wissenschaften und Künste,* vol. 42, edited by Johann Heinrich Zedler, 377–82. Leipzig: Zedler, 1744.

Thomasius, Christian. *Versuch vom Wesen des Geistes, oder Grund-Lehren, so wohl zur natürlichen Wissenschaft, als der Sitten-Lehre.* Halle: Salfeld, 1699.

Tschirnhaus, Ehrenfried Walther von. *Medicina Mentis Sive Tentamen genuinae Logicae, in quâ disseritur De Methodo detegendi incognitas veritates.* Amsterdam: Apud Albertum Magnum, and Joannem Rieuwerts Juniorem, 1687. Reprinted as *Medicina Mentis et Corporis,* Leipzig: Fritsch, 1695.

———. "Brief an Leibniz" [Kieslingswalde den 13. Januar 1693]. In *Sämtliche Schriften und Briefe Gottfried Wilhelm Leibniz,* no. 1, *Allgemeiner, Politischer und Historischer Briefwechsel,* vol. 9, *1693,* edited by Kurt Müller, Günter Scheel and Gerda Utermöhlen, 463–67. Berlin: Akademie der Wissenschaften, 1975 and 1992.

———. "Brief an A. H. Francke" [January 17, 1698]. Reprinted in "August Hermann Francke und Ehrenfried Walter von Tschirnhaus: Eine Bekanntschaft im Spiegel der Quellen im Achiv der Franckeschen Stiftungen Halle/Saale," by Mathias Ullmann, in *Um Gott zu Ehren und zu des Landes Besten: Die Franckeschen Stiftungen und Preussen; Aspekte einer alten Allianz*, edited by Thomas Müller-Bahlke, 318–19. Halle: Franckeschen Stiftungen, 2001.

———. *Gründliche Anleitung zu nüzlichen Wissenschafften absonderlich zu der Mathesi und Physica. Wie sie anitzo von den Gelehrtesten abhandelt werden.* Frankfurt and Leipzig, 1708.

Uffenbach, Zacharias Conrad. *Merkwüridge Reisen durch Niedersachsen, Holland und Engelland* [1709]. Part 2. Ulm: Gaum, 1753.

Voltaire. *Philosophical Dictionary*. 2 vols. Translated and edited by Peter Gay. New York: Basic Books, 1962.

Weigel, Erhard. "Kurtzer Entwurff der freudigen Kunst- und Tugend-Lehr, vor Trivial und Kinder Schulen." In *Gesammelte pädagogische Schriften*, edited by Hermann Schüling, 64–72. Giessen: Universitätsbibliothek, 1970.

Weise, Christian. *Das erste Jahr-Gedächtnüs der wohlgehaltenen Bet-Woche, das ist: die wiederhohlte Dancksagung des neuerbauten und numehr in Stand gebrachten Waysenhauses in Zittau.* Zittau, 1702.

Whiston, William. *A New Theory of the Earth from its Original to the Consummation of All Things.* London: Tooke, 1696.

———. *Sir Isaac Newton's Mathematick Philosophy more easily demonstrated with Dr. Halley's account of comets illustrated.* London, 1716.

———. *The Longitude and Latitude Found by the Inclinatory or Dipping Needle: wherein the Laws of Magnetism are also dicover'd.* London: 1719, 1721.

Wöchenliche Hallische Frage- und Anzeigungsnachrichten. Halle, 1729–31.

Wöchenliche Hallische Anzeigen. Halle, 1731–1809.

Wolff, Christian. *Anfangsgründe aller mathematischen Wissenschaften.* Halle: Renger, 1710.

———. *Mathematisches Lexicon, darinnen die in allen Theilen der Mathematick üblichen Kunst-Wörter erkläret, und zur Historie der Mathematischen Wissenschafften dienliche Nachrichten ertheilet, auch die Schrifften, wo jede Materie ausgeführet zu finden.* Leipzig: Gleditsch, 1716.

———. *Vernünftige Gedanken von Gott, der Welt und der Seele des Menschen, auch allen Dingen überhaupt.* Halle: Renger, 1720. Fourth edition [1751] reprinted as "Vernünftige Gedanken," in *Gesammelte Werke*, I, vol. 2, edited by Jean École and Charles Corr. Hildesheim: Olms, 2004.

———. *Allerhand Nützliche Versuche, dadurch zu genauer Erkäntnis der Natur und Kunst der Weg gebähnhet.* Halle: Renger, 1721–23.

———. *Oratio de Sinarum Philosophia Practica.* Frankfurt am Main: Andreae and Hort, 1726. Reprinted as *Rede über die praktische Philosophie der Chinesen*, edited and translated by Michael Albrecht. Hamburg: Felix Meiner, 1985.

———. *Eigene Lebensbeschreibung*. Edited by Heinrich Wuttke. Leipzig: Weidmann-sche Buchhandlung, 1841.

———. *Biographie*. With a preface by Hans Werner Arndt. Hildesheim: Olms, 1980.

———. "Gutachten über einige vom Pastor Eberhard erfundene nautische Instru-mente" [June 15, 1720]. In *Briefe von Christian Wolff aus den Jahren 1719–1753: Ein Beitrag zur Geschichte der Kaiserlichen Academie der Wissenschaften zu St. Petersburg*, edited by v. Kunitz, translated from Russian into German by Bärbel Vinz, doc. no. 152, p. 293. St. Petersburg: Eggens, 1860.

———. *Philosophia Rationalis Sive Logica . . . Praemittitur Discursus Praeliminaris De Philsophia In Genere* [1728]. Reprinted as *Preliminary Discourse on Philosophy in General*, translated and edited by Richard Blackwell. Indianapolis: Bobbs-Merrill, 1963.

Zahn, Gottfried. *Erste Nachricht von einer auf Gr. Königlichen Majestät allergnädigste Concession angefangenen Waisen und Schul Anstalt zu Bunzlau in Schlesien Welche sich auf den Fond der Göttlichen Vorsehung gründet. Mit beigefügten Gedanken von dergleichen Unternehmungen*. Bunzlau: Waisenhaus, 1754.

Secondary Sources

Ahnert, Thomas. *Religion and the Origins of the German Enlightenment: Faith and the Reform of Learning in the Thought of Christian Thomasius*. Rochester, NY: University of Rochester Press, 2006.

Albrecht, Michael. *Eklektik: Eine Begriffsgeschichte mit Hinweisen auf die Philosophie- und Wissenschaftsgeschichte*. Stuttgart: Frommann-Holzboog, 1994.

———. "Hypothesen und Phänomene: Zu Sturms Theorie der wissenschaftliche Methode." In *Acta Historica Astronomiae*, vol. 22, *Johann Christoph Sturm, 1635–1703*, edited by Hans Gaab, Pierre Leich, and Günther Löffladt, 119–35. Frankfurt: Harri Deutsch, 2004.

Albrecht-Birkner, Veronika. *Reformation des Lebens: Die reformen Herzog Ernst des Frommen von Sachsen-Gotha und ihre Auswirkungen auf Frömmigkeit, Schule und Alltag im ländlichen Raum (1640–1675)*. Leipzig: Evangelische, 2002.

———. *Hoffnung besserer Zeiten: Philipp Jakob Spener (1635–1705) und die Geschichte des Pietismus*. Halle: Franckeschen Stiftungen, 2005.

Andrews, William J. H., ed. *The Quest for Longitude: The Proceedings of the Longitude Symposium*. Cambridge, MA: Collection of Scientific Instruments, Harvard University, 1996.

Antognazza, Maria Rosa. *Leibniz: An Intellectual Biography*. Cambridge: Cambridge University Press, 2009.

Antognazza, Maria Rosa, and Howard Hotson. *Alsted and Leibniz: On God, the Magistrate and the Millennium*. Wiesbaden: Harrassowitz, 1999.

Arnold, Ken. "Trade, Travel and Treasure: Seventeenth-Century Artificial Curiosi-ties." in *Transports: Travel, Pleasure, and Imaginative Geography, 1600–1830*, edited by Chloe Chard and Helen Langdon, 263–85. New Haven, CT: Yale University Press, 1996.

Baird, David. *Thing Knowledge: A Philosophy of Scientific Instruments.* Berkeley: University of California Press, 2004.

Baird, Olga. "I Want the People to Observe and to Learn! The St. Petersburg *Kunstkamera* in the Eighteenth Century." *History of Education* 37 (July 2008): 531–47.

Baldwin, Martha R. "Magnetism and the Anti-Copernican Polemic." *Journal for the History of Astronomy* 16 (1985): 155–74.

———. "Athanasius Kircher and the Magnetic Philosophy." PhD diss., University of Chicago, 1987.

Bates, Robert, and Julia Jackson, eds. *Glossary of Geology.* Alexandria, VA: American Geological Institute, 1980.

Baumgart, Peter. "Leibniz und der Pietismus." *Archiv für Kulturgeschichte* 48 (1966): 364–86.

Becker, George. "Pietism's Confrontation with Enlightenment Rationalism: An Examination of the Relation between Ascetic Protestantism and Science." *Journal for the Scientific Study of Religion* 30 (June 1991): 139–58.

Bennett, James A. "The Challenge of Practical Mathematics." In *Science, Culture and Popular Belief in Renaissance Europe*, edited by Stephen Pumfrey, Paolo L. Rossi, and Maurice Slawinski, 176–90. Manchester: Manchester University Press, 1991.

Bennett, James A., and Scott Mandelbrote. *The Garden, the Ark, the Tower, the Temple: Biblical Metaphors of Knowledge in Early Modern Europe.* Oxford: Museum of the History of Science, 1998.

Bernet, Claus. *"Gebaute Apokalypse": Die Utopie des Himmlischen Jerusalem in der Frühen Neuzeit. Veröffentlichungen des Instituts für Europäische Geschichte Mainz: Abteilung für Abendländische Religiongeschichte.* Mainz: Philipp von Zabern, 2007.

Bethmann, Louis. "Geschichte der Anstalt." In *Landwaisen-Anstalt Langendorf: Festschrift zur Feier ihres 200 jährigen Bestehens*, 7–74. Langendorf, 1910.

Beutel, Albrecht. "Causa Wolffiana: Die Vertreibung Christian Wolffs aus Preussen 1723 als Kulminationspunkt des theologisch-politischen Konflikts zwischen halleschem Pietismus und Aufklärungsphilosophie." in *Wissenschaftliche Theologie und Kirchenleitung*, edited by Ulrich Köpf, 159–202. Tübingen: Mohr Siebeck, 2001.

Beyreuther, Erich. *Geschichte des Pietismus.* Stuttgart: Steinkopf, 1978.

Biagioli, Mario. *Galileo, Courtier: The Practice of Science in the Culture of Absolutism.* Chicago: University of Chicago Press, 1994.

Biesler, Jorg. *BauKunstKritik: Deutsche Architekturtheorie im 18.Jahrhundert.* Berlin: Dietrich Remer Verlag, 2005.

Blair, Ann. "Mosaic Physics and the Search for a Pious Natural Philosophy in the Late Renaissance." *Isis* 91 (March 2000): 32–58.

Bleichmar, Daniela. *Visible Empire: Botanical Expeditions and Visual Culture in the Hispanic Enlightenment.* Chicago: University of Chicago Press, 2012.

Bleichmar, Daniela, and Peter Mancall, eds. *Collecting across Cultures: Material Exchanges in the Early Modern Atlantic World.* Philadelphia: University of Pennsylvania Press, 2011.

Bloth, Hugo. *Johann Julius Hecker (1707–1768) und seine Universalschule.* Dortmund: Crüwell, 1968.

Boros, Gábor, Herman De Dijn, and Martin Moors, eds. *The Concept of Love in 17th and 18th Century Philosophy.* Leuven, Belg.: Leuven University Press, 2007.

Bos, Jacques. "Individuality and Inwardness in the Literary Character Sketches of the Seventeenth Century." *Journal of the Warburg and Courtauld Institutes* 61 (1998): 142–57.

Brather, Hans-Stephan, ed. *Leibniz und seine Akademie: Ausgewählte Quellen zur Geschichte der Berliner Sozietät der Wissenschaften, 1697–1716.* Berlin: Akademie, 1993.

Brecht, Martin. "August Hermann Francke und der Hallische Pietismus." In *Geschichte des Pietismus,* vol. 1, edited by Klaus Depperman, Hartmut Lehmann, and Ulrich Gäbler, 440–61. Göttingen: Vandenhoeck and Ruprecht, 1993.

———. "Probleme der Pietismusforschung." *Dutch Review of Church History* 76 (1997): 227–37.

Bredekamp, Horst. *The Lure of Antiquity and the Cult of the Machine: The Kunstkammer and the Evolution of Nature, Art, and Technology.* Translated by Allison Brown. Princeton, NJ: Markus Wiener, 1995.

———. "Leibniz's Theater of Nature and Art and the Idea of a Universal Picture Atlas." In *The Artificial and the Natural: An Evolving Polarity,* edited by Bensaude-Vincent and William R. Newman, 211–23. Cambridge, MA: MIT Press, 2007.

Breul, Wolfgang, Marcus Meier, and Lothar Vogel, eds. *Der radikale Pietismus: Perspektiven der Forschung.* Göttingen: Vandenhoeck and Ruprecht, 2010.

Brockliss, Laurence. "Curricula: The Faculty of Law." In *Universities in Early Modern Europe, 1500–1800,* edited by Hilde de Ridder-Symoen, 599–609. Cambridge: Cambridge University Press, 1996.

Bruning, Jens. *Innovation in Forschung und Lehre: die Philosophische Fakultät der Universität Helmstedt in der Frühaufklärung 1680–1740.* Wiesbaden: Harrassowitz, 2012.

Brunner, Daniel L. *Halle Pietists in England: Anthony William Böhm and the Society for Promoting Christian Knowledge.* Göttingen: Vandenhoeck and Ruprecht, 1993.

Bryden, D. J. "Magnetic Inclinatory Needles: Approved by the Royal Society?" *Notes and Records of the Royal Society of London* 47 (January 1993): 17–31.

Buchenau, Stephanie. *The Founding of Aesthetics in the German Enlightenment: The Art of Invention and the Invention of Art.* Cambridge: Cambridge University Press, 2013.

Busch, Ralf. "Johan Lund, seine Alten Jüdischen Heiligtümer und die Vorstellung vom Salomonischen Tempel." *Jewish Art* 19–20 (1994): 62–67.

Camille, Michael. "Before the Gaze: The Internal Senses and Late Medieval Practices of Seeing." In *Visuality before and beyond the Renaissance: Seeing as Others*

Saw, edited by Robert Nelson, 197–223. Cambridge: Cambridge University Press, 2000.

Carpo, Mario. "How Do You Imitate a Building That You Have Never Seen? Printed Images, Ancient Models, and Handmade Drawings in Renaissance Architectural Theory." *Zeitschrift für Kunstgeschichte* 64 (2001): 223–33.

Cervenka, Jaromir. *Die Naturphilosophie des Johann Amos Comenius*. Prague: Academia, 1970.

Chadarevian, Soraya de, and Nick Hopwood, eds. *Models: The Third Dimension of Science*. Stanford, CA: Stanford University Press, 2004.

Chorpenning, Joseph, ed. *Emblemata Sacra: Emblem Books from the Maurits Sabbe Library, Katholieke Universiteit Leuven*. Philadelphia: Saint Joseph's Unviersity Press, 2006.

Clark, Christopher. *The Iron Kingdom: The Rise and Downfall of Prussia, 1600–1947*. Cambridge, MA: Harvard University Press, 2007.

Clark, Stuart. *Vanities of the Eye: Vision in Early Modern Culture*. Oxford: Oxford University Press, 2007.

Clark, William. *Academic Charisma and the Origins of the Research University*. Chicago: University of Chicago Press, 2006.

Classen, Constance. *Worlds of Sense: Exploring the Senses in History and across Cultures*. New York: Routledge, 1993.

Cohen, Claudine, and Andre Wakefield. Introduction to *Protogaea*, by Gottfried Wilhelm Leibniz, xiii–xl. Chicago: University of Chicago Press, 2008.

Cohen, Mark A. "La Bruyère and the 'Usage' of Childhood: The Idea of Pedagogy in the Caractères." *French Forum* 26 (Spring 2001): 23–42.

Coleman, Charly. "Resacralizing the World: The Fate of Secularization in Enlightenment Historiography." *Journal of Modern History* 82 (June 2010): 368–95.

Collis, Robert. *The Petrine Instauration: Religion, Esotericism and Science at the Court of Peter the Great, 1689–1729*. Leiden: Brill, 2011.

Conrads, Norbert. *Ritterakademien der Frühen Neuzeit: Bildung als Standesprivileg im 16. und 17. Jahrhundert*. Vandenhoeck and Ruprecht, 1982.

Cook, Alan. "Edmund Halley and the Magnetic Field of the Earth." *Notes and Records of the Royal Society of London* 55 (September 2000): 473–90.

Cook, Harold. *Matters of Exchange: Commerce, Medicine, and Science in the Dutch Golden Age*. New Haven, CT: Yale University Press, 2007.

Cooper, Alix. *Inventing the Indigenous: Local Knowledge and Natural History in Early Modern Europe*. Cambridge: Cambridge University Press, 2007.

Cracraft, James. *The Petrine Revolution in Russian Culture*. Cambridge, MA: Belknap, 2004.

Cross, Anthony Glenn. *Peter the Great through British Eyes: Perceptions and Representations of the Tsar since 1698*. Cambridge: Cambridge University Press, 2000.

Cunningham, Hugh, and Joanna Ines, eds. *Charity, Philanthropy, and Reform from the 1690s to 1850*. Basingstoke: Macmillan; New York: St. Martin's, 1998.

Daston, Lorraine. "On Scientific Observation." *Isis* 99 (2008): 97–110.

———. "The Empire of Observation, 1600–1800." In *Histories of Scientific Observa-*

tion, edited by Lorraine Daston and Elizabeth Lunbeck, 81–113. Chicago: University of Chicago Press, 2011.

Daston, Lorraine, and Elizabeth Lunbeck, eds. *Histories of Scientific Observation*. Chicago: University of Chicago Press, 2011.

Davison, Lee, Tim Hitchcock, Tim Keim, and Robert B. Shoemaker, eds. *Stilling the Grumbling Hive: The Response to Social and Economic Problems in England, 1689–1750*. New York: St. Martin's, 1992.

Deppermann, Klaus. *Der hallesche Pietismus und der preußische Staat unter Friedrich III*. Göttingen: Vandenhoeck and Ruprecht, 1961.

Dickson, Donald R., "Johann Valentin Andreae's Utopian Brotherhoods." *Renaissance Quarterly* 49 (1996): 760–802.

———. *The Tessera of Antilia: Utopian Brotherhoods and Secret Societies in the Early Seventeenth Century*. Leiden: Brill, 1998.

Dijksterhuis, Fokko Jan. "Moving around the Ellipse: Conic Sections in Leiden, 1620–1660." In *Silent Messengers: the Circulation of Material Objects of Knowledge in the Early Modern Low Countries*, edited by Sven Dupre and Christopher Lüthy, 89–124. Berlin: Lit. Verlag, 2011.

Dolezel, Eva. "Inszenierte Objekte: Der Indienschrank in der Kunst und Naturalienkammer der Franckeschen Stiftugen zu Halle." In *Fremde Dinge*, edited by Michael C. Frank et al., 29–38. Bielefeld: transcript-Verl 2007.

Downy, Glanville. "Philanthropia in Religion and Statecraft in the Fourth Century after Christ." *Historia: Zeitschrift für Alte Geschichte* 4:2–3 (1955): 199–208.

Drechsler, Wolfgang. "Christian Wolff (1679–1754): A Biographical Essay." In *The European Journal of Law and Economics* 4 (1997): 111–28.

Eiskildsen, Kasper Risbjerg. "Christian Thomasius, Invisible Philosophers, and Education for Enlightenment." *Intellectual History Review* 18 (2008): 319–36.

Eliav-Feldon, Miriam. *Realistic Utopias: The Ideal Imaginary Societies of the Renaissance, 1516–1630*. Oxford: Clarendon, 1982.

Eller, David B. "The Recovery of the Love Feast in German Pietism." In *Confessionalism and Pietism: Religious Reform in Early Modern Europe*, edited by Fred van Lieburg, 11–30. Mainz: Philipp von Zabern, 2006.

Evans, R. J. "Learned Societies in Germany in the Seventeenth Century." *European Studies Review* 7 (1977): 129–51.

Evers, Bernd, ed. *Architekturmodelle der Renaissance. Die Harmonie des Bauens von Alberti bis Michelangelo*. Munich: Prestel, 1996.

Fara, Patricia. *Sympathetic Attractions: Magnetic Practices, Beliefs, and Symbolism in Eighteenth-Century England*. Princeton, NJ: Princeton University Press, 1996.

———. "Hidden Depths: Halley, Hell and Other People." *Studies in History and Philosophy of Science* 38 (2007): 570–83.

Farrell, Maureen. *William Whiston*. New York: Arno, 1981.

Fehr, James Jakob. *"Ein wunderlicher Nexus Rerum": Aufklärung und Pietismus in Königsberg unter Franz Albert Schultz*. Hildesheim: Georg Olms, 2005.

Feingold, Mordechai. "Tradition versus Novelty: Universities and Scientific

Societies in the Early Modern Period." In *Revolution and Continuity: Essays in the History and Philosophy of Early Modern Science*, edited by Peter Barker and Roger Ariew, 45–62. Washington, DC: Catholic University of America Press, 1991.

Ferngren, Gary B., ed. *Science and Religion: A Historical Introduction*. Baltimore: Johns Hopkins University Press, 2002.

Findlen, Paula. *Possessing Nature: Museums, Collecting, and Scientific Culture in Early Modern Italy*. Berkeley: University of California Press, 1994.

———. "Inventing Nature: Commerce, Art and Science in the Early Modern Cabinet of Curiosities." In *Merchants and Marvels: Commerce, Science and Art in Early Modern Europe*, edited by Pamela Smith and Paula Findlen, 297–323. New York: Routledge, 2002.

Fischer, Kuno. "Leibnizens kirchenpolitische Wirksamkeit: die Unionsbestrebungen." In *Gottfried Wilhelm Leibniz: Leben, Werke und Lehre*, edited by T. S. Hoffmann, 165–72. Wiesbaden: Marix, 1920/2009.

Fitting, Peter, ed. *Subterranean Worlds: A Critical Anthology*. Middletown, CT: Wesleyan University Press, 2004.

Force, James. *William Whiston, Honest Newtonian*. Cambridge: Cambridge University Press, 1985.

Friedrich, Karin. *Brandenburg-Prussia, 1466–1806*. Basingstoke: Palgrave Macmillan, 2012.

Frost, Robert. *The Northern Wars: War, State, and Society in Northeastern Europe, 1558–1721*. Essex: Pearson, 2000.

Fulbrook, Mary. *Piety and Politics: Religion and the Rise of Absolutism in England, Württemburg, and Prussia*. Cambridge: Cambridge University Press, 1983.

Futch, Michael. "Leibniz on Plenitude, Infinity and the Eternity of the World." *British Journal for the History of Philosophy* 10 (2002): 541–60.

Gascoigne, John. "A Reappraisal of the Role of the Universities in the Scientific Revolution." In *Reappraisals of the Scientific Revolution*, edited by David C. Lindberg and Robert S. Westman, 207–60. Cambridge: Cambridge University Press, 1990.

Gaukroger, Stephen. "Descartes' Early Doctrine of Clear and Distinct Ideas." *Journal of the History of Ideas* 53 (1992): 585–602.

———. *Francis Bacon and the Transformation of Early Modern Philosophy*. Cambridge: Cambridge University Press, 2001.

Gawthrop, Richard L. *Pietism and the Making of Eighteenth-Century Prussia*. Cambridge: Cambridge University Press, 1993.

———. "Pedagogy and Reform in Halle Pietism and the German Enlightenment." In *Interdisziplinäre Pietismusforschungen*, edited by Udo Sträter et al., 529–36. Halle: Max Niemeyer, 2005.

Gay, Peter. *The Enlightenment: An Interpretation*. Vol. 1, *The Rise of Modern Paganism*. New York: W. W. Norton, 1966.

Gerber, Georg. "Die Neu-Atlantis des Francis Bacon und die Entstehung der Academia Naturae Curiosorum (Leopoldina) und der Societät der Wissenschaften in Berlin." In *Wissenschaftliche Annalen*, edited by Hans Wittbrodt, 552–60. Berlin: Akademie, 1955.

Geyer-Kordesch, Johanna. *Pietismus, Medizin und Aufklärung in Preussen im 18.Jahrhundert: Das Leben und Werk Georg Ernst Stahls*. Tübingen: Niemeyer, 2000.

Gierl, Martin. *Pietismus und Äufklärung: Theologische Polemik und die Kommunikationsreform der Wissenschaft am Ende des 17.Jahrhunderts*. Göttingen: Vandenhoeck and Ruprecht, 1997.

———. "Im Netz der Theologen—Die Wiedergeburt der Geschichte findet nicht statt: Von Pietismusforschung, protestantischer Identität und historischer Ethik." *Zeitschrift für Historische Forschung* 32 (2005): 463–87.

Gleixner, Ulrike. "Expansive Frömmigkeit: Das hallische Netzwerk der Indienmission im 18.Jahrhundert." In *Mission und Foschung: Translokale Wissensproduktion zwischen Indien und Europa im 18. & 19.Jahrhundert*, edited by Heike Liebau et al., 57–66. Halle: Franckeschen Stiftungen, 2010.

Goldgar, Anne. *Impolite Learning: Conduct and Community in the Republic of Letters, 1680–1750*. New Haven, CT: Yale University Press, 1995.

Golinski, Jan. "A Noble Spectacle: Phosphorus and the Public Cultures of Science in the Early Royal Society." *Isis* 80 (1989): 11–39.

Goodrick-Clarke, Clare. "The Rosicrucian Afterglow: The Life and Influence of Comenius." In *The Rosicrucian Enlightenment Revisited*, by John Matthews et al., with an introduction by Ralph White, 93–218. Hudson, NY: Lindisfarne Books, 1999.

Gordin, Michael. "The Importation of Being Earnest: The Early St. Petersburg Academy of Sciences." *Isis* 91 (2000): 1–31.

Gorman, Michael. "The Angel and the Compass: Athanasius Kircher's Magnetic Geography." In *Athanasius Kircher: The Last Man Who Knew Everything*, edited by Paula Findlen, 239–59. New York: Routledge, 2004.

Gorsky, Phil. *The Disciplinary Revolution: Calvinism and the Rise of the State in Early Modern Europe*. Chicago: University of Chicago Press, 2003.

Gowing, Laura, Michael Hunter, and Miri Rubin, eds. *Love, Friendship and Faith in Europe, 1300–1800*. Basingstoke: Palgrave Macmillan, 2005.

Grau, Conrad. *Die Preußische Akademie der Wissenschaften zu Berlin*. Heidelberg: Spektrum Akademischer, 1993.

Greengrass, Mark, Michael Leslie, and Timothy Raylor, eds. *Samuel Hartlib and Universal Reformation: Studies in Intellectual Communication*. Cambridge: Cambridge University Press, 1994.

Gross, Andreas, ed. *Halle and the Beginning of Protestant Christianity in India*. Vol. 1, *The Danish-Halle and the English-Halle Mission*. Halle: Franckesche Stiftungen, 2006.

Grote, Simon. "Pietistische Aisthesis und moralische Erziehung bei Alexander Gottlieb Baumgarten." *Aufklärung* 20 (2008): 107–27.

———. "Moral Philosophy and the Origins of Modern Aesthetic Theory." PhD diss., University of California, Berkeley, 2010.

———. "Review-Essay: Religion and Enlightenment." *Journal of the History of Ideas* 75 (2014): 137–60.

Guerrier, W. *Leibniz in seinen Beziehungen zu Russland und Peter dem Grossen.* St. Petersburg and Leipzig, 1873.

Haakonssen, Knud. *Natural Law and Moral Philosophy: From Grotius to the Scottish Enlightenment.* Cambridge: Cambridge University Press, 1996.

Hadot, Pierre. *Philosophy as a Way of Life: Spiritual Exercises from Socrates to Foucault.* Edited by Arnold Davidson, translated by Michael Chase. Oxford: Blackwell, 1995.

———. *What Is Ancient Philosophy?* Translated by Michael Chase. Cambridge, MA: Belknap, 2002.

Hall, A. Rupert, and Marie Boas Hall, eds. *The Correspondence of Henry Oldenburg.* Vol. 7, *1670–1671.* Madison: University of Wisconsin Press, 1970.

Hall, Marie Boas. "Slare, F. R. S. (1648–1727)." *Notes and Records of the Royal Society of London* 46 (January 1992): 23–41.

Hamblin, William J., and David Rolph Seely. *Solomon's Temple: Myth and History.* London: Thames and Hudson, 2007.

Hamm, Ernst P. "Knowledge from Underground: Leibniz Mines the Enlightenment." *Earth Sciences History* 16 (1997): 77–99.

Harries, Karsten. *The Ethical Function of Architecture.* Cambridge, MA: MIT, 1998.

Harrington, Joel. *The Unwanted Child: The Fate of Foundlings, Orphans, and Juvenile Criminals in Early Modern Germany.* Chicago: University of Chicago Press 2009.

Harris, Steven J. "Confession-Building, Long-Distance Networks and the Organization of Jesuit Science." *Early Science and Medicine* 1 (1996): 287–318.

Härter, Carl. "Sozialdisziplinierung." In *Frühe Neuzeit,* edited by Annette Völker-Rasor, 294–98. Munich: R. Oldenbourg, 2000.

Hartkopf, Werner. *Die Berliner Akademie der Wissenschaften: Ihre Mitglieder und Preisträger 1700–1990.* Berlin: Akademie, 1992.

Headley, John M. *Tommoso Campanella and the Transformation of the World.* Princeton, NJ: Princeton University Press, 1997.

Helm, Jürgen, and Elisabeth Quast, eds. *Fromme Unternehmer: Briefe der Ärzte Christian Friedrich und Christian Sigismund Richer an Carl Hildebrand v. Canstein.* Halle: Franckesche Stiftungen, 2010.

Hettche, Matt. "On the Cusp of Europe's Enlightenment: Christian Wolff and the Argument for Academic Freedom." *Florida Philosophical Review* 8 (2008): 91–107.

Heyd, Michael. *"Be Sober and Reasonable": The Critique of Enthusiasm in the Seventeenth and Early Eighteenth Centuries.* Leiden: Brill, 1995.

Hine, William. "Athanasius Kircher and Magnetism." In *Athanasius Kircher und seine Beziehungen zum gelehrten Europa seiner Zeit,* Wolfenbütteler Arbeiten zur Barockforschung 17, edited by John Fletcher, 79–97. Wiesbaden: Harrassowitz, 1998.

Hinrichs, Carl. *Preußentum und Pietismus: Der Pietismus in Brandenburg-Preußen als religiös-soziale Reformbewegung.* Göttingen: Vandenhoeck and Ruprecht, 1971.

Hochstrasser, Timothy. *Natural Law Theories in the Early Enlightenment*. Cambridge: Cambridge University Press, 2000.

Hofmann, Franz. "Consultatio Catholica de emendatione rerum humanarum: Gedanken zur Edition der halleschen Funde des Spätwerks des J. A. Comenius durch die Tschechoslowakische Akademie der Wissenschaften." *Wissenschaftliche Zeitschrift der Martin-Luther-Universität Halle- Wittenberg. Gesellschafts- und sprachwissenschaftliche*, no. 17 (1968): 127–49.

———. "Die 'Halleschen Funde'—Schicksal einer Handschrift." *Mitteilungsbl. d. Comeniusforschungsstelle Bochum* 25 (1992): 39–55.

Holloran, John Robert. "Professors of Enlightenment at the University of Halle, 1690–1730." PhD diss., University of Virginia, 2000.

Hopf, Friedrich Wilhelm. "Anton, Paul." In *Neue Deutsche Biographie*, 1:319–20. Berlin: Duncker and Humblot, 1953.

Hotson, Howard. "Irenicism and Dogmatics in the Confessional Age: Pareus and Comenius in Heidelberg, 1614." *Journal of Ecclesiastical History* 45 (July 1995): 432–53.

———. "Irenicism in the Confessional Age: The Holy Roman Empire, 1563–1648." In Conciliation and Confession: The Struggle for Unity in the Age of Reform, 1415–1648, edited by Howard P. Louthan and Randall C. Zachman, 228–85. Notre Dame, IN: University of Notre Dame Press, 2004.

Houston, Chloe. "'Knowledge Shall Be Increased': Natural Philosophy and Religion in the Early Modern Utopia." *Literature Compass* 4 (2007): 1397–1411.

Howard, Thomas Albert. *Protestant Theology and the Making of the Modern German University*. Oxford: Oxford University Press, 2006.

Howarth, R. J. "Fitting Geomagnetic Fields before the Invention of Least Squares: II. William Whiston's Isoclinic Maps of Southern England (1719 and 1721)." *Annals of Science* 60 (2003): 63–84.

Hsia, Florence. "Mathematical Martyrs, Mandarin Missionaries, and Apostolic Academicians: Telling Institutional Lives." In *Institutional Culture in Early Modern Society*, edited by Anne Goldgar and Robert I. Frost, 3–34. Leiden: Brill, 2004.

———. *Sojourners in a Strange Land: Jesuits and Their Scientific Missions in Late Imperial China*. Chicago: University of Chicago Press, 2009.

Hsia, Ronnie Po-Chia. *Social Discipline in the Reformation: Central Europe, 1550–1750*. New York: Routledge, 1989.

Hughes, Thomas P. "Model Builders and Instrument Makers." *Science in Context* 2 (1986): 59–75.

Hunter, Ian. "The Love of a Sage or the Command of a Superior: The Natural Law Doctrines of Leibniz and Pufendorf." In *Early Modern Natural Law Theories: Contexts and Strategies in the Early Enlightenment*, edited by T. J. Hochstrasser and P. Schröder, 169–94. Dordrecht: Kluwer Academic, 2003.

Hunter, Michael. *Boyle: Between God and Science*. New Haven, CT: Yale University Press, 2009.

Impey, Oliver, and Arthur MacGregor, eds. *The Origins of Museums: The Cabinet of*

Curiosities in Sixteenth- and Seventeenth-Century Europe. Oxford: Clarendon, 1985.

Israel, Jonathan I. *The Radical Enlightenment: Philosophy and the Making of Modernity, 1650–1750.* Oxford: Oxford University Press, 2001.

———. *Enlightenment Contested: Philosophy, Modernity, and the Emancipation of Man, 1670–1752.* New York: Oxford University Press, 2006.

Jacobi, Juliane, ed. *"Man hatte von ihm gute Hoffnung": Das Waisenalbum der Franckeschen Stiftungen, 1695–1749.* Halle: Franckeschen Stiftungen im Niemeyer-Verlag Tübingen, 1998.

Jakubowski-Tiessen, Manfred. "Der Pietismus in Dänemark und Schleswig Holstein." In *Geschichte des Pietismus,* vol. 2, edited by Hartmut Lehmann, 446–70. Göttingen: Vandenhoeck and Ruprecht, 2003.

"Johann Daniel Herrnschmidt." In *Allgemeine Deutsche Biographie,* 12:221–22. Leipzig: Duncker and Humblot, 1880.

Johns, Adrian. *The Nature of the Book: Print and Knowledge in the Making.* Chicago: University of Chicago Press, 1998.

Jones, Matthew L. *The Good Life in the Scientific Revolution: Descartes, Pascal, Leibniz and the Cultivation of Virtue.* Chicago: University of Chicago Press, 2006.

Jordan, George Jeff. *The Reunion of the Churches: A Study of G. W. Leibniz and His Great Attempt.* London: Constable, 1927.

Jung, Volker. *Das Ganze der Heiligen Schrift: Hermeneutik und Schriftauslegung bei Abraham Calov.* Stuttgart: Calver Verlag, 1999.

Kanthak, Gerhard. *Der Akademiegedanke zwischen utopischem Entwurf und barocker Projektmacherei.* Berlin: Dunker and Humblot, 1987.

Karant-Nunn, Susan. *The Reformation of Feeling: Shaping the Religious Emotions in Early Modern Germany.* Oxford: Oxford University Press, 2010.

Kaulbach, Friedrich. "Anschauug." In *Historisches Wörterbuch der Philosophie,* vol. 1, edited by Joachim Ritter, 340–47. Basel: Schwabe, 1971.

Kelley, Donald. "Eclecticism and the History of Ideas." *Journal of the History of Ideas* 62 (2001): 577–92.

Kevorkian, Tanya. "Piety Confronts Politics: Philipp Jacob Spener in Dresden, 1686–1691." *German History* 16:2 (1998): 145–46.

King, Lester S. "Stahl and Hoffmann: A Study in Eighteenth-Century Animism." *Journal of the History of Medicine and Allied Sciences* 19 (1964): 118–30.

Klein, Ursula, and E. C. Spary, eds. *Materials and Expertise in Early Modern Europe: Between Market and Economy.* Chicago: University of Chicago Press, 2010.

Kleinert, Andreas. "Johann Joachim Lange (1699–1765), ein unbekannter Hallenser Mathematikprofessor im Schatten von Christian Wolff." *Acta Historica Leopoldina* 54 (2008): 477–88.

Klemme, Heiner. *Die Schule Immanuel Kants: Mit dem Text Christian Schiffert über das Königsberger Collegium Fridericianum.* Hamburg: Meiner, 1994.

Klüger, Richard. "Die pädagogischen Ansichten des Philosophen Tschirnhaus." PhD diss., University of Leipzig, 1913.

Kohlenberger, Helmut K. "Anschauung Gottes." In *Historisches Wörterbuch der Philosophie*, vol. 1, edited by Joachim Ritter, 347–49. Basel: Schwabe, 1971.

Kollerstrom, Nicholas. "The Hollow World of Edmund Halley." *Journal of the History of Astronomy* 23 (1992): 185–92.

Kramer, Gustav. *August Hermann Francke's Pädagogische Schriften*. Langesalza: Beyer, 1876.

———. *August Hermann Francke's Pädagogische Schriften*. Langensalza: Beyer, 1885.

———. *August Hermann Francke: Ein Lebensbild*. Parts 1 and 2. Halle: Buchhandlung des Waisenhauses. 1880–82. Reprint, Hildesheim: Georg Olms, 2004.

Kühnel, Johannes. *Comenius und der Anschauungsunterricht*. Leipzig: Klinkhardt, 1911.

LaVopa, Anthony. *Grace, Talent, and Merit: Poor Students, Clerical Careers, and Professional Ideology in Eighteenth-Century Germany*. Cambridge: Cambridge University Press, 1988.

Lawlor, Robert. *Sacred Geometry: Philosophy and Practice*. New York: Thames and Hudson, 1989.

Lehmann, D. Arno. *Alte Briefe aus Indien: Unveröffentlichte Briefe von Bartholomäus Ziegenbalg, 1706–1719*. Berlin: Evangelische, 1957.

Lehmann, Hartmut. "Enger, weiterer und erweiterer Pietismusbegriff. Anmerkungen zu den kirtischen Anfragen von Johannes Wallmann an die Konzeption der Geschichte des Pietismus." *Pietismus und Neuzeit* 29 (2003): 18–36.

———. "Erledigte und nichte erledigte Aufgaben der Pietismusforschung: eine nochmalige Antwort an Johannes Wallmann." *Pietismus und Neuzeit* 31 (2005): 13–20.

Lempa, Heike, and Paul Peucker, eds. *Self, Community, World: Moravian Education in a Transatlantic World*. Bethlehem, PA: Lehigh University Press, 2010.

Leventhal, Robert. "Enlightenment Uncontested." Review of *Enlightenment Contested: Philosophy, Modernity and the Emancipation of Man, 1670–1752*, by Jonathan Israel. H-German, H-Net Reviews, June 2007.

Liebmann, Otto. "Walter von Tschirnhaus." In *Allgemeine Deutsche Biographie*, vol. 38 (1894): 722–24.

Lindberg, Carter. *The Pietist Theologians: An Introduction to Theology in the Seventeenth and Eighteenth Centuries*. Oxford: Blackwell, 2005.

Lipski, Alexander. "The Foundation of the Russian Academy of Sciences." *Isis* 44 (1953): 349–54.

Lowood, Henry. *Patriotism, Profit, and the Promotion of Science in the German Enlightenment: The Economic and Scientific Societies, 1760–1815*. New York: Garland, 1991.

Lux, David. "Societies, Circles, Academies and Organizations: A Historiographic Essay on Seventeenth-Century Science." In *Revolution and Continuity: Essays in the History and Philosophy Early Modern Science*, edited by Peter Barker and Roger Ariew, 23–43. Washington, DC: Catholic University of America Press, 1991.

Lynch, Michael. "The Production of Scientific Images: Vision and Re-Vision in the

History, Philosophy and Sociology of Science." In *Visual Cultures of Science: Rethinking Representational Practices in Knowledge Building and Science Communication*, edited by Luc Pauwels, 26–40. Hanover, NH: Dartmouth College Press, University Press of New England, 2006.

Lynch, William T. *Solomon's Child: Method in the Early Royal Society of London*. Stanford, CA: Stanford University Press, 2001.

MacClellan, James. *Science Reorganized: Scientific Societies in the Eighteenth Century*. New York: Columbia University Press, 1988.

Mager, Inge. *Georg Calixts theologische Ethik und ihre Nachwirkungen*. Göttingen: Vandenhoeck and Ruprecht, 1969.

Mahnke, Dietrich. "Der Barock Universalismus des Comenius." In *Zeitschrift für Geschichte der Erziehung und des Unterrichts*, 21:96–128. Langensalza, 1931.

Marshall, John. *John Locke, Toleration, and Early Enlightenment Culture: Religious Intolerance and the Arguments for Religious Toleration in Early Modern and "Early Enlightenment" Europe*. Cambridge: Cambridge University Press, 2006.

Marschke, Benjamin. *Absolutely Pietist: Patronage, Factionalism and State-Building in the Early Eighteenth-Century Army Chaplaincy*. Tübingen: Franckeschen Stiftungen Halle im Max Niemeyer, 2005.

———. "Lutheran Jesuits: Halle Pietist Communication Networks at the Court of Frederick William I of Prussia." *Covenant Quarterly* 65:4 (2006): 19–38.

———. "Wir Hallenser: The Understanding of Insiders and Outsiders among Halle Pietists in Prussia under Frederick William I (1713–1740)." In *Pietism and Community in Europe and North America, 1650–1850*, edited by Jonathan Strom, 81–94. Leiden: Brill, 2010.

Masser, Karin. *Christóbal de Gentil de Rojas y Spinola O.F.M. und der lutherische Abt Gerardus Wolterius Molanus: Ein Beitrag zur Geschichte der Unionsbestrebungen der katholischen und evangelischen Kirche im 17.Jahrhundert*. Münster: Aschendorff, 2002.

Mauss, Marcel. *The Gift: The Form and Reason for Exchange in Archaic Societies*. Trans. Mary Douglas. London: Routledge, 2001.

Mayer, Uwe. "Am Rande der Gelehrtenrepublik—Tschirnhaus als Mathematiker." In *Ehrenfried Walther von Tschirnhaus (1651–1708): Experimente mit dem Sonnenfeuer; Sonderausstellung im Mathematischen-Physikalischen Salon im Dresdner Zwinger vom 11. April 2001 bis 29.Juli 2001*, edited by Peter Plaßmeyer, 25–35. Dresden: Staatliche Kunstsammlungen, 2001.

McClellan, James, III. *Science Reorganized: Scientific Societies in the Eighteenth Century*. New York: Columbia University Press, 1985.

McClure, Edmund, ed. *A Chapter in English Church History: Being the Minutes of the Society for Promoting Christian Knowledge for the Years 1698–1704*. London, 1888.

Melton, James Van Horn. *Absolutism and the Eighteenth-Century Origins of Compulsory Schooling in Prussia and Austria*. Cambridge: Cambridge University Press, 1988.

———. "Pietism and the Public Sphere in Eighteenth-Century Germany." In

Religion and Politics in Enlightenment Europe, edited by James E. Bradley and Dale K. Van Kley, 294–333. Notre Dame, IN: University of Notre Dame Press, 2001.

———. *The Rise of the Public in Enlightenment Europe*. Cambridge: Cambridge University Press, 2001.

Menck, Peter. *Die Erziehung der Jugend zur Ehre Gottes und zum Nutzen des Nächsten: Die Pädagogik August Hermann Franckes*. Halle: Franckesche Stiftungen; Tübingen: Max Niemeyer, 2001.

Mentzel, Friedrich-Franz. *Pietismus und Schule: Die Auswirkungen des Pietismus auf das Berliner Schulwesen, 1691–1797*. Hohengehren: Schneider, 1994.

———. "Ein erfolgreicher Pietist an König Friedrichs Hof? Der Briefwechsel von Johann Julius Hecker und Gotthilf August Francke (1746–1763)." *Berliner Aufklärung* 1 (1999): 13–40.

Mercer, Christia. "Platonism and Philosophical Humanism on the Continent." In *A Companion to Early Modern Philosophy*, edited by Steven Nadler, 25–44. Oxford: Blackwell, 2002.

Merkel, Francis R. "Missionary Attitude of the Philosopher Gottfried Wilhelm Leibniz." In *Christianity and Missions, 1450–1800*, edited by J. S. Cummins, 291–302. Brookfield, VT: Ashgate, 1997.

Merkel, Franz Rudolf. *G. W. von Leibniz und die China-Mission*. Leipzig: Hinrichs, 1920.

Merrill, Ronald T. *Our Magnetic Earth: The Science of Geomagnetism*. Chicago: University of Chicago Press, 2010.

Merton, Robert K. *Social Theory and Social Structure*. Glencoe, IL: Free Press, 1957.

———. "The Fallacy of the Latest Word: The Case of Pietism and Science." *American Journal of Sociology* 89 (March 1984): 1091–1121.

Mettele, Gisela. "Identities across Borders: The Moravian Brethren as a Global Community." In *Pietism and Community in Europe and North America, 1650–1850*, edited by Jonathan Strom, 155–77. Boston: Brill, 2010.

Meyer, R. W. *Leibniz and the Seventeenth-Century Revolution*. Translated by J. P. Stern. Cambridge: Bowes and Bowes, 1952.

Michael, Gerhard. *Die Welt als Schule: Ratke, Comenius und die didaktische Bewegung*. Hannover. Schrocdcl, 1978.

Morgan, Mary S., and Margaret Morrison, eds. *Models as Mediators: Perspectives on Natural and Social Science*. Cambridge: Cambridge University Press, 1999.

Mosley, Adam. "Objects of Knowledge: Mathematics and Models in Sixteenth-Century Cosmology and Astronomy." In *Transmitting Knowledge: Words, Images, and Instruments in Early Modern Europe*, edited by Sachiko Kusukawa and Ian Maclean, 193–216. Oxford: Oxford University Press, 2006.

Mühlpfordt, Günter. "Halle—Russland—Sibirien—Amerika: Georg Wilhlem Steller, der Hallesche Kolumbus und Halles Anteil an der frühen Osteuropa und Nordasienforshchung." In *Hallesche Forschung*, vol. 1, *Halle und Osteuropa zur europäische Ausstrahlung des hallischen Pietismus*, edited by J. Wallmann and Udo Sträter, 49–82. Tübingen: Max Niemyer, 1998.

Müller-Bahlke, Thomas. "Der Realienunterricht in den Schulen August Hermann Francke." In *Schulen Machen Geschichte: 300 Jahre Erziehung in den Franckeschen Stiftungen zu Halle*, edited by Carmela Keller and Thomas Müller-Bahlke, 43–65. Halle: Franckeschen Stiftungen, 1997.

———. *Die Wunderkammer. Die Kunst- und Naturalienkammer der Franckeschen Stiftungen zu Halle (Saale)*. Halle: Franckeschen Stiftungen, 1998.

———, ed. *Um Gott zu Ehren und zu des Landes Besten: Die Franckeschen Stiftungen und Preußen; Aspekte einer alten Allianz*. Halle: Franckesche Stiftungen, 2001.

———. "Die frühe Verwaltungsstrukturen der Franckeschen Stiftungen." In *Waisenhäuser in der Frühen Neuzeit*, edited by Udo Sträter, Joseph N. Neumann, and Renate Wilson, 41–70. Halle: Franckeschen Stiftungen, 2003.

———. "Naturwissenschaft und Technik: Der Hallesche Pietismus am Vorabend der Industrialisierung." In *Geschichte des Pietismus*, vol. 4, *Glaubenswelt und Lebenswelten*, edited by Hartmut Lehmann, 357–85. Göttingen: Vandenhoeck and Ruprecht, 2003.

Mulsow, Martin. *Die drei Ringe: Toleranz und clandestine Gelehrsamkeit bei Mathurin Veyssière La Croze (1661–1739)*. Tübingen: Niemeyer, 2001.

Murphy, Daniel. *Comenius: A Critical Re-Assessment of His Life and Work*. Dublin: Irish Academic, 1995.

Naragon, Steve. "Friedrich Hoffmann." In *The Dictionary of Eighteenth-Century German Philosophers*, vol. 1, edited by Manfred Kuehn and Heiner Klemme, 526–29. London: Continuum, 2010.

Nischan, Bodo. "John Bergius: Irenicism and the Beginning of Official Religious Toleration in Brandenburg-Prussia." *Church History* 51 (1982): 389–404.

Nummendal, Tara. *Alchemy and Authority in the Holy Roman Empire*. Chicago: University of Chicago Press, 2007.

Obst, Helmut. *August Hermann Francke und die Franckesche Stiftungen*. Göttingen: Vandenhoeck and Ruprecht, 2002.

Offenburg, A. K. "Jacob Jehuda Leon (1602–1675) and His Model of the Temple." In *Jewish-Christian Relations in the Seventeenth Century: Studies and Documents*, edited by J. van den Berg and E. G. van der Wall, 95–115. Dordrecht: Kluwer, 1988.

Olgilvie, Brian. "Natural History, Ethics, and Physico-Theology." In *Historia: Empiricism and Erudition in Early Modern Europe*, edited by Gianna Pomata and Nancy G. Siraisi, 75–103. Cambridge, MA: MIT Press, 2005.

Ornstein, Martha. *The Role of Scientific Societies in the Seventeenth Century*. Chicago: University of Chicago Press, 1928.

Osler, Margaret. "Mixing Metaphors: Science and Religion or Natural Philosophy and Theology in Early Modern Europe." *History of Science* 36 (1998): 91–113.

———. *Reconfiguring the World: Nature, God, and Human Understanding from the Middle Ages to Early Modern Europe*. Baltimore: Johns Hopkins University Press, 2010.

Outram, Dorinda. *The Enlightenment*. Cambridge: Cambridge University Press, 2005.

Pedersen, Olaf. "The Rise of the Academies." In *Universities in Early Modern Europe, 1500–1800*, edited by Hilde de Ridder-Symoens, 480–88. Cambridge: Cambridge University Press, 1996.

Peschke, Erhard. "Die Reformideen des Comenius und ihr Verhältnis zu A. H. Francke's Plan einer realen Verbesserung in der ganzen Welt." In *Die Pietismus in Gestalten und Wirkungen*, edited by Heinrich Bornkamm, Friedrich Heyer, and Alfred Schindler, 368–83. Bielefeld: Luther, 1975.

Phillips, Denise. *Acolytes of Nature: Defining Natural Science in Germany, 1770–1850*. Chicago: University of Chicago Press, 2012.

Pott, Sandra, Martin Mulsow, and Lutz Danneberg, eds. *The Berlin Refuge, 1680–1780:* Learning and Science in European Context. Leiden: Brill, 2003.

Pozzo, Richard. "Prejudices and Horizons: G. F. Meier's *Vernunftlehre* and Its Relation to Kant." *Journal of the History of Philosophy* 43 (2005): 185–202.

Pumfrey, Stephan. "Mechanizing Magnetism in Restoration England—the Decline of Magnetic Philosophy." *Annals of Science* 44 (1987): 1–22.

———. "Magnetical Philosophy and Astronomy, 1600–1650." In *Planetary Astronomy from the Renaissance to the Rise of Astrophysics*, Part A, *Tycho Brahe to Newton*, edited by Rene Taton and Curtis Wilson, 45–53. Cambridge: Cambridge University Press, 1989.

———. *Latitude and the Magnetic Earth: The True Story of Queen Elizabeth's Most Distinguished Man of Science*. Cambridge: Icon Books, 2002.

Quast, Elizabeth. "Schwägerinnen: Adlige Frauen in der Frühphase der Halleschen Medikamentenexpedition." In *Medical Theory and Therapeutic Practice in the Eighteenth Century: A Transatlantic Perspective*, edited by Jürgen Helm and Renate Wilson, 281–307. Stuttgart: Fritz Steiner, 2008.

Raeff, Marc. *The Well-Ordered Police State: Social and Institutional Change through Law in the Germanies and Russia, 1600–1800*. New Haven, CT: Yale University Press, 1983.

Rajan, Kamath. "Der Beitrag der Dänisch-Halleschen Missionare zum europäischen Wissen über Indien im 18.Jahrhundert." In *Mission und Foschung: Translokale Wissensproduktion zwischen Indien und Europa im 18. & 19.Jahrhundert*, edited by Heike Liebau et al., 93–112. Halle: Franckeschen Stiftungen, 2010.

Ramati, Ayval. "Harmony at a Distance: Leibniz's Scientific Academies." *Isis* 87 (1996): 430–52.

Ramirez Jasso, Diana. "Imagining the Garden: Childhood, Landscape and Architecture in Early Pedagogy, 1761–1850." PhD diss., Harvard University, 2012.

Rappaport, Rhoda. *When Geologists Were Historians, 1665–1750*. Ithaca, NY: Cornell University Press, 1997.

Raspopov, O. M., and V. V. Meshcheryakov. "Magnetic Declination Measurements over European Russia and Siberia in the 18th Century." *Geomagnetism and Aeronomy* 51 (2011): 1146–54.

Reill, Peter Hanns. *Vitalizing Nature in the Enlightenment*. Berkeley: University of California Press, 2003.

Reimbold, Ernst Thomas. *Pia Desideria Gottselige Begierden*. Olten: Walter, 1980.

Reinhard, Wolfgang. "Zwang zur Konfessionalisierung? Prolegomena zu einer Theorie des konfessionallen Zeitalters." *Zeitschrift für historische Forschung* 10 (1983): 257–77.

Rescher, Nicholas. "Leibniz Visits Vienna (1712–1714)." *Studia Leibnitiana* 31 (1999): 133–59.

Rich, Michael. "Representing Euclid in the Eighteenth Century." In *The Whipple Museum of the History of Science: Instruments and Interpretations* . . . , by Liba Chaia Taub and Frances Willmoth, 319–44. Cambridge: Cambridge University Press, 2006.

Riley, Patrick. *Leibniz's Universal Jurisprudence: Justice as the Charity of the Wise*. Cambridge, MA: Harvard University Press, 1996.

———. "Leibniz's Political and Moral Philosophy in the *Novissima Sinica*, 1699–1999." *Journal of the History of Ideas* 60 (1999): 217–39.

Ritschl, Albrecht. *Geschichte des Pietismus*.3 vols. Bonn: Marcus, 1880–86.

Roberts, Lissa. "Mapping Steam Engines and Skill in Eighteenth-Century Holland." In *The Mindful Hand: Inquiry and Invention from the Late Renaissance to Early Industrialisation*, edited by Lissa Roberts, Simon Schaffer, and Peter Dear, 197–220. Amsterdam: Koninklijke Nederlandse Akademie van Wetenschappen, 2007.

———. "Situating Science in Global History: Local Exchanges and Networks of Circulation." *Itinerario* 33 (2009): 9–30.

Roberts, Lissa, Simon Schaffer, and Peter Dear, eds. *The Mindful Hand: Inquiry and Invention from the Late Renaissance to Early Industrialisation*. Amsterdam: Koninklijke Nederlandse Akademie van Wetenschappen, 2007.

Robinson, Robert, and Walter Adams, eds. *The Diary of Robert Hooke, 1672–1680*. London: Taylor and Francis, 1935.

Rosenau, Helen. *Vision of the Temple: The Image of the Temple of Jerusalem in Judaism and Christianity*. London: Oresko Books, 1979.

Rossi, Paulo. *The Dark Abyss of Time: The History of the Earth & the History of Nations from Hooke to Vico*. Translated by Lydia G. Cochrane. Chicago: University of Chicago Press, 1984.

———. *Logic and the Art of Memory: The Quest for a Universal Language*. Translated with an introduction by Stephen Cuclas. Chicago: University of Chicago Press, 2000.

Rule, Paul. "François Noel, SJ, and the Chinese Rites Controversy." In *The History of the Relations between the Low Countries and China in the Qing Era (1644–1911)*, edited by Willy Vande Walle and Noël Golvers, 137–66. Leuven: Leuven University Press, 2003.

Ryan, W. F. "Scientific Instruments in Russia from the Middle Ages to Peter the Great." *Annals of Science* 48 (1991): 367–84.

Rykwert, Joseph. *On Adam's House in Paradise: The Idea of the Primitive Hut in Architectural History*. New York: Museum of Modert Art, 1972.

Safley, Thomas Max. *Charity and Economy in the Orphanages of Early Modern Augsburg*. Atlantic Highlands, NJ: Humanities Press International, 1997.

———. *Children of the Laboring Poor: Expectation and Experience among the Orphans of Early Modern Augsburg*. Leiden: Brill, 2005.

Schadel, Erwin. *Sehendes Herz (cor oculatum)-zu einem Emblem des späten Comenius: Prämdoernes Seinsverständnis als Impuls für integral konzipierte Postmoderne* Frankfurt: Peter Lang, 2003.

Schaffer, Simon. "Halley's Atheism and the End of the World." *Notes and Records of the Royal Society of London* 32 (1977): 17–40.

———. "Machine Philosophy: Demonstration Devices in Georgian Mechanics." In "Instruments," edited by Albert van Helden and T. L. Hankins, special issue, *Osiris* 9 (1994): 157–82.

———. "The Show That Never Ends: Perpetual Motion in the Early Eighteenth Century." *British Journal for the History of Science* 28 (1995): 157–89.

Schaper, Joachim. "The Jerusalem Temple as an Instrument of the Achaemenid Fiscal Adminstration." *Vetus Testamentum* 45 (1995): 528–39.

Schechner, Sara J. *Comets, Popular Culture, and the Birth of Modern Cosmology*. Princeton, NJ: Princeton University Press, 1996.

Schicketanz, Peter, ed. *Carl Hildebrand v. Cansteins Beziehungen zu Philipp Jakob Spener*. Witten: Luther, 1967.

———. *Der Briefwechsel Carl Hildebrand von Cansteins mit August Hermann Francke*. Berlin: Walter de Gruyter, 1972.

Schickore, Jutta. *The Microscope and the Eye: A History of Reflections, 1740–1870*. Chicago: University of Chicago Press, 2007.

Schille, Christa. "Christoph Eberhard." In *Neue deutsche Biographie*, 4:238–39. Berlin: Duncker and Humblot, 1959.

Schilling, Heinz. *Religion, Political Culture, and the Emergence of Early Modern Society: Essays in German and Dutch History*. Leiden: Brill, 1992.

Schlagenhauf, Wilfried. "Ansätze einer technikbezogenen Bildung in Schulkonzepten um 1700." In *Technikvermittlung und Technikpopularisierung: Historische und didaktische Perspektiven*, edited by Lars Bluma, Karl Pichol, and Wolfhard Weber, 197–211. Münster: Waxmann, 2004.

Schlüsslcr, Hermann. *Georg Calixt, Theologie und Kirchen politik*. Wiesbaden: Steiner, 1961.

Schmidt, Hanno. "On the Importance of Halle in the Eighteenth Century for the History of Education." *Paedagogica Historica* 32 (1996): 85–100.

Schmidt-Biggemann, Wilhelm. "Die Historisierung der 'Philosophia Hebraeorum' im frühen 18.Jh." In *Aporemata: Kritische Studien zur Philologiegeschichte*, vol. 5, *Historicization-Historisierung*, edited by Glenn Most, 103–28. Göttingen: Vandenhoeck and Ruprecht, 2001.

Schmoldt, Benno, and Michael Soren Schuppan. *Materialien und Studien zur Geschichte der Berliner Schule*. Goeppingen: Schneider, 1993.

Schneewind, Jerome B. "Philosophical Ideas of Charity: Some Historical Reflections." In *Giving: Western Ideas of Philanthropy*, edited by Jerome B. Schnee-wind, 54–75. Bloomington: Indiana University Press, 1996.

Schneider, Hans. *German Radical Pietism*. Translated by Gerald T. MacDonald. Lanham, MD: Scarecrow, 2007.

Schneider, Ulrich Johannes. "Eclecticism and the History of Philosophy." In *History of the Disciplines: The Reclassification of Knowledge in Early Modern Europe*, edited by Donald Kelley, 83–101. Rochester, NY: University of Rochester Press, 1997.

Schrader, Wilhelm. *Geschichte der Friedrichs Universität zu Halle*. 2 vols. Berlin: Dümmler, 1894.

Schunka, Alexander. "Daniel Ernst Jablonski, Pietism, and Ecclesiastical Union." In *Pietism, Revivalism and Modernity, 1650–1850*, edited by Fred van Lieburg and Daniel Lindmark, 23–41. Newcastle upon Tyne: Cambridge Scholars Publishing, 2008.

———. "Zwischen Kontingenz und Providenz: Frühe Englandkontakte der halleschen Pietisten und protestantische Irenik um 1700." *Pietismus und Neuzeit* 34 (2008): 82–114.

Scott, John Beldon. *Architecture for the Shroud: Relic and Ritual in Turin*. Chicago: University of Chicago Press, 2003.

Scribner, Robert. "Ways of Seeing in the Age of Dürer." In *Dürer and His Culture*, edited by Dagmar Eichberger and Charles Zika, 93–117. Cambridge: Cambridge University Press, 1998.

Shantz, Douglas. *Between Sardis and Philadelphia: The Life and World of Pietist Court Preacher Conrad Bröske*. Leiden: Brill, 2008.

———. "Conversion and Revival in the Last Days: Hopes for Progress and Renewal in Radical Pietism and Gottfried Wilhelm Leibniz." In *Pietism, Revivalism and Modernity 1650–1850*, edited by Fred van Lieburg and Daniel Lindmark, 242–62. Newcastle upon Tyne: Cambridge Scholars Publishing, 2008.

———. *An Introduction to German Pietism: Protestant Renewal at the Dawn of Modern Europe*. Young Center Books in Anabaptist and Pietist Studies. Baltimore: Johns Hopkins University Press, 2013.

Shapin, Steven. "The House of Experiment in Seventeenth-Century England." *Isis* 79 (1988): 373–404.

———. "Robert Boyle and Mathematics: Reality, Representation, and Experimental Practice." *Science in Context* 2 (1988): 23–58.

———. *A Social History of Truth: Civility and Science in Seventeenth-Century England*. Chicago: University of Chicago Press, 1994.

———. *The Scientific Revolution*. Chicago: University of Chicago Press, 1996.

Shapin, Steven, and Simon Schaffer. *Leviathan and the Air-Pump: Hobbes, Boyle, and the Experimental Life*. Princeton, NJ: Princeton University Press, 1985.

Sheehan, Jonathan. "Enlightenment, Religion, and the Enigma of Secularization: A Review Essay." *American Historical Review* 108 (2003): 1061–80.

———. *The Enlightenment Bible: Translation, Scholarship, Culture*. Princeton, NJ: Princeton University Press, 2005.

———. "Temple and Tabernacle: The Place of Religion in Early Modern England." In *Making Knowledge in Early Modern Europe: Practices, Objects, and Texts, 1400–1800*, edited by Pamela H. Smith and Benjamin Schmidt, 248–341. Chicago: University of Chicago Press, 2007.

Sibum, Otto. "Experimentalists in the Republic of Letters." *Science in Context* 16 (2003): 89–120.

Smith, Pamela. *The Business of Alchemy: Science and Culture in the Holy Roman Empire*. Princeton, NJ: Princeton University Press, 1994.

———. *The Body of the Artisan: Art and Experience in the Scientific Revolution*. Chicago: University of Chicago Press, 2004.

Smith, Pamela, and Paula Findlen, eds. *Merchants and Marvels: Commerce, Science and Art in Early Modern Europe*. New York: Routledge, 2002.

Snobelen, S. D. "William Whiston: Natural Philosopher, Prophet, Primitive Christian." PhD diss., University of Cambridge, 2000.

Snobelen, S. D., and Larry Stewart. "Making Newton Easy: William Whiston in Cambridge and London." In *From Newton to Hawking: a History of Cambridge University's Lucasian Professors of Mathematics*, edited by Kevin Knox and Richard Noakes, 135–70. Cambridge: Cambridge University Press, 2003.

Sobel, Dava. *Longitude: The True Story of a Lone Genius Who Solved the Greatest Scientific Problem of His Time*. New York: Walker, 1995.

Sorkin, David Jan. *The Religious Enlightenment: Protestants, Jews, and Catholics from London to Vienna*. Princeton, NJ: Princeton University Press, 2008.

Spaans, Joke. "Early Modern Orphanages between Civic Pride and Social Discipline: Francke's Use of Dutch Models." In *Waisenhäuser der Frühen Neuzeit*, edited by Udo Sträter, 183–96. Tübingen: Nieymer, 2003.

Sparn, Walter. "Philosophie." In *Geschichte des Pietismus*, vol. 4, edited by Hartmut Lehmann, 227–63. Göttingen: Vandenhoeck and Ruprecht, 2003.

Stewart, Larry. *The Rise of Public Science: Rhetoric, Technology, and Natural Philosophy in Newtonian Britain, 1660–1750*. Cambridge: Cambridge University Press, 1992.

———. "Other Centres of Calculation; or, Where the Royal Society Didn't Count: Commerce, Coffee-Houses and Natural Philosophy in Early Modern London." *British Journal for the History of Science* 32 (1999): 133–53.

Stitziel, Judd. "God, the Devil, Medicine, and the Word: A Controversy over Ecstatic Women in Protestant Middle Germany, 1691–1693." *Central European History* 29 (1996): 309–37.

Stolzenburg, W. A. H. *Geschichte des Bunzlauer Waisenhaus*. Vol. 3. Bunzlau: Waisenhaus, 1851.

Sträter, Udo. "Pietismus und Sozialtätigkeit: Zur Frage nach der Wirkungsgeschichte des 'Waisenhauses' in Halle und des Frankfurter Armen-, Waisen- und Arbeitshauses." *Pietismus und Neuzeit* 8 (1982): 201–30.

———. *Sonthom, Bayly, Dyke and Hall: Studien zur Rezeption der englischen Erbauungsliteratur in Deutschland im 17.Jahrhundert*. Tübingen: Mohr, 1987.

———. *Meditation und Kirchenreform in der lutherische Kirche des 17.Jahrhunderts*. Tübingen: C. B. Mohr, 1995.

———. "Wie bringen wir den Kopf in das Herz?" In *Meditation und Erinnerung in der Frühe Neuzeit*, edited by Gerhard Kurz, 11–35. Göttingen: Vandenhoeck and Ruprecht, 2000.

———. "Zum Verhältnis des frühen Pietismus zu den Naturwissenschaften." *Pietismus und Neuzeit* 32 (2005): 79–100.

Strom, Jonathan. "Problems and Promises of Pietism Research." *Church History* 71 (September 2002): 536–54.

———, ed. *Pietism and Community in Europe and North America*. Leiden: Brill, 2010.

Sulek, Marty. "On the Classical Meaning of Philanthropia." *Nonprofit and Voluntary Sector Quarterly* 39 (2010): 385–408.

Summers, David. *Real Spaces: World Art History and the Rise of Western Modernism*. New York: Phaidon 2003.

Sutton, Geoffrey V. *Science for a Polite Society: Gender, Culture, and the Demonstration of Enlightenment*. Boulder, CO: Westview, 1997.

Te Heesen, Anke. "Boxes in Nature." *Studies in History and Philosophy of Science*. Part A, 30 (September 2000): 381–403.

———. *The World in a Box: The Story of an Eighteenth-Century Picture Encyclopedia*. Chicago: University of Chicago Press, 2001.

Trepp, Anne-Charlotte. *Von der Glückseligkeit alles zu wissen: Die Erforschung der Natur als religiöse Praxisin in der Frühen Neuzeit (1550–1750)*. Frankfurt am Main: Campus, 2009.

Turnbull, G. H. *Hartlib, Dury and Comenius: Gleanings from Hartlib's Papers*. Liverpool: University Press of Liverpool, 1947.

Ulbricht, Otto. "Foundling Hospitals in Enlightenment Germany: Infanticide, Illegitimacy and Infant Mortality Rates." *Central European History* 18 (1985): 215.

Ullmann, Mathias. "August Hermann Francke und Ehrenfried Walter von Tschirnhaus: Eine Bekanntschaft im Spiegel der Quellen im Achiv der Franckeschen Stiftungen Halle / Saale." In *Um Gott zu Ehren und zu des Landes Besten. Die Franckeschen Stiftungen und Preussen: Aspekte einer alten Allianz*, edited by Thomas Müller-Bahlke, 317–34. Halle: Franckeschen Stiftungen, 2001.

Utermöhlen, Gerda. "Die Russlandthematik im Briefwechsel zwischen August Hermann Francke und Gottfried Wilhelm Leibniz." In *Hallesche Forschung*, vol. 1, *Halle und Osteuropa zur europäische Ausstrahlung des hallischen Pietismus*, edited by Johannes Wallmann and Udo Sträter, 109–28. Tübingen: Max Niemyer, 1998.

Van Hoorn, Tanja, and Yvonne Wübben, eds. *"Allerhand nützliche Versuche": Empirische Wissenskultur in Halle und Göttingen (1720–1750)*. Hannover: Wehrhahn, 2009.

Van Peursen, C. A. "E. W. von Tschirnhaus and the Ars Inveniendi." *Journal of the History of Ideas* 54 (1993): 395–410.

Van Vaeck, Marc, and John Mannings, eds. *The Jesuits and the Emblem Tradition: Selected Papers of the 4th Leuven International Emblem Conference, 18–23 August, 1996*. Turnhout: Brepols, 1999.

Vial, Alexander. *Dr. Conrad Mel, weiland geistl: Inspector und Rector des Gymnasium zu Hersfeld; Ein Lebensbild aus dem Ende des XVII und Anfang des XVIII Jahrhunderts*. Hersfeld, 1864.

Vidal, Fernando. "Introduction: Knowledge, Belief and the Impulse to Natural Theology." *Science in Context* 20 (2007): 381–400.

———. *The Sciences of the Soul: The Early Modern Origins of Psychology*. Translated by Saskia Brown. Chicago: University of Chicago Press, 2011.

Vierhaus, Rudolf. "The Prussian Bureaucracy Reconsidered." In *Rethinking Leviathan: The Eighteenth-Century State in Britain and Germany*, edited by John Brewer and Eckhart Hellmuth, 150–64. London: German Historical Institute, 1999.

Vogelsang, Bernd. "'Archaische Utopien': Materialien zu Gerhard Schotts Hamburger 'Bühnenmodell' des Templum Salomonis." PhD diss., University of Cologne, 1981.

Vollrath, Hans-Joachim. "Das Pantometrum Kircherianum—Athanasius Kirchers Messtisch." In *Spurensuche—Wege zu Athanasius Kircher*, edited by H. Beinlich, H. J. Vollrath, and K. Wittstadt, 119–36. Dettelbach: Röll, 2002.

Von Naredi-Rainer, Paul. *Salomos Tempel und das Abendland: Monumentale Folgen Historischer Irrtümer*. Cologne: Dumont Buchverlag 1994.

Wakefield, Andre. *The Disordered Police State: German Cameralism as Science and Practice*. Chicago: University of Chicago Press, 2009.

Wallmann, Johannes. *Philipp Jakob Spener und die Anfänge des Pietismus*. 2nd rev. ed. Tübingen: Mohr, 1986.

———. "Was ist Pietismus?" *Pietismus und Neuzeit* 21 (1994): 220–31.

———. "Union, Reunion, Toleranz: Georg Calixts Einigungsbestrebungen und ihre Rezeption in der Katholischen und Protestantischen Theologie des 17.Jahrhunderts." In *Union-Konversion-Toleranz: Dimension der Annährung zwischen den christlichen Konfessionen im 17. und 18.Jahrhundert*, edited by Heinz Duchhardt and Gerhard May, 331–48. Mainz: Philipp von Zabern, 2000.

———. "Eine alternative Geschichte des Pietismus: zur gegenwärtigen Diskussion um den Pietismusbegriff." *Pietismus und Neuzeit* 28 (2002): 30–71.

———. "Pietismus—ein Empochenbegriff oder ein typoloigscher Begriff? Antwort auf Hartmut Lehmann." *Pietismus und Neuzeit* 30 (2004): 191–224.

Ward, W. Reginald. *The Protestant Evangelical Awakening*. Cambridge: Cambridge University Press, 1992.

Warner, Deborah. "What Is a Scientific Instrument? When Did It Become One and Why?" *British Journal for the History of Science* 23 (1990): 83–93.

———. "Terrestrial Magnetism: For the Glory of God and the Benefit of Mankind." In "Instruments," edited by Albert van Helden and T. L. Hankins, special issue, *Osiris* 9 (1994): 67–84.

Webster, Charles. *The Great Instauration: Science, Medicine, and Reform, 1626–1660*. New York: Homes and Meier, 1975.

Wendebourg, Dorothea, ed. *Philipp Jakob Spener—Leben, Werk, Bedeutung: Bilanz der Forschung nach 300 Jahren.* Tübingen: Franckeschen Stiftungen; Halle: Max Niemeyer, 2007.

Whitmer, Kelly J. "Eclecticism and the Technologies of Discernment in Pietist Pedagogy." *Journal of the History of Ideas* 70 (2009): 545–67.

———. "Unmittelbare Erkenntnis. Das Modell des Salomonische Tempels im Waisenhaus zu Halle als Anschauungsobjekt der frühen Aufklärung." *Bildwelten des Wissens: Kunsthistorisches Jahrbuch für Bildkritik* 7:1 (2009): 92–104.

———. "The Model That Never Moved: The Case of a Virtual Memory Theater and Its Christian Philosophical Argument." *Science in Context* 23 (2010): 289–327.

———. "What's in a Name? Place, Peoples and Plants in the Danish-Halle Mission, c. 1710–1740." *Annals of Science* 23 (2013): 289–327.

Wilson, Catherine. "De Ispa Natura: Leibniz's Doctrine of Force, Activity and Natural Law," *Studia Leibnitiana* 19 (1987): 148–72.

Wilson, Renate. "Heinrich Wilhelm Ludolf, August Hermann Francke und der Eingang nach Russland." In *Hallesche Forschung*, vol. 1, *Halle und Osteuropa zur europäischen Ausstrahlung des hallischen Pietismus*, edited by Johannes Wallmann and Udo Sträter, 83–108. Halle: Franckesche Stiftungen, 1998.

———. "Philanthropy in Eighteenth-Century Central Europe: Evangelical Reform and Commerce." *Voluntas: International Journal of Voluntary and Nonprofit Organizations* 9 (March 1998): 81–102.

———. *Pious Traders in Medicine: A German Pharmaceutical Network in Eighteenth-Century North America.* University Park: Pennsylvania State University Press, 2000.

———. "Replication Reconsidered: Imitations, Models and the Seeds of Modern Philanthropy." In *Waisenhäuser in der Fruehen Neuzeit*, by Udo Sträter and Josef N. Neumann. Halle: Franckeschen Stiftungen, 2003.

Winter, Edward. *Halle als Ausgangspunkt der deutschen Russlandkunde im 18.Jahrhundert.* Berlin: Deutsche Akadmie der Wissenschaft, 1953.

Wiseman, Boris, ed. *The Cambridge Companion to Lévi-Strauss.* Cambridge: Cambridge University Press, 2009.

Wollgast, Siegfried. *E. W. v. Tschirnhaus und die deutsche Frühaufklärung.* Berlin: Akademie, 1988.

Wotschke, Theodor. "Das pietistische Halle und die Auslandsdeutschen." *Neue kirchliche Zeitschrift* 43 (1932): 428–34, 475–92.

Yates, Francis A. *The Rosicrucian Enlightenment.* London: Routledge and Kegan Paul, 1972.

Zeeden, Ernst Walter. *Die Entstehung der Konfessionen: Grundlagen und Formen der Konfessionsbildung.* Munich: R. Oldenbourg, 1965.

Zippel, Gustav. *Geschichte des Königlichen Friedrichs-Kollegium zu Königsberg Pr. 1698–1898.* Königsberg, 1898.

Zytaruk, Maria. "Cabinets of Curiosities and the Organization of Knowledge." *University of Toronto Quarterly* 80 (Winter 2011): 1–23.

Index

drawing lessons in the Orphanage, 84; and model making, 6, 80–82
Friedrich Wilhelm I: and irenical turn, 43–44; on observing the Orphanage, 1; as soldier king, 147n33; tour of the Orphanage with Francke, 4–7, 9, 87, 106; and Wolff controversy, 13

geography: Eberhard's son as tutor of, 106; in Halle Orphanage and *Pädagogium*, viii, 5, 56, 58; in Hecker's *Realschule*, 119–20; Jesuits as teachers of, 93; in Königsberg, 121, 123; in Langendorf Orphanage, 113; recommended by Francke for Tobolsk Orphanage, 125; as taught by Pietist teachers, 91
geomagnetism, 92, 95, 103–4
globes: crafting of, 63–64, 125, 162n67; magnetic, viii, 86, 93–94, 98–99

Halle Orphanage: cameralism and, 9; as model, 5–6, 31; and newsletter of (*die Hallesche Berichte*), 160; role of teacher in, 13–14; as philanthropic organization, 9; as seminar of nations, 4, 30; as showplace (*Schauplatz*), 1; as universal facility, 3, 10; University of Halle and, 3, 11
Halley, Edmund, 93–94, 99–100, 156n19, 157nn40–41, 158n58
Hecker, Johann Julius, 109, 118–20, 130
Herrnschmidt, Johann Daniel: and eclecticism, 50–52, 127; observation, 13, 51–52; as student of Sturm, 50, 149n68
history: of the Earth, 54, 102; in Hecker's *Realschule*, 119; in Königsberg, 121; and mosaic physics, 54–55, 58; natural, 14, 17, 29; and observation of models, 76; of the Orphanage, 108, of philosophy, 54; recommended by Francke for Tobolsk Orphanage, 125; as subject taught in Halle Orphanage and *Pädagogium*, viii, 5, 38, 56, 123; and theology, 54, 125
Hoffmann, Friedrich, viii, 12, 118, 138n47
hollow earth, 93–94, 99–100, 159n68
Huguenots, 40

imaginables, 26–27, 62, 127
imagination, 25–27, 79
inclination. *See* magnetic inclination
inclinator (*instrumentum inclinatorium*). *See under* instruments

inner eye. *See* observation(s): Anschauung
instrumental philosophy. *See under* philosophy
instruments: air pump, 6, 30, 51, 120, 126; camera obscura, 82, 84; compass, 93; crafting of, 6; and education, 91; in Hecker's *Realschule*, 120; inclinator, viii, 19, 87, 92–107, 126, 128, 158nn58–59, 158n63; in Leibniz's ideal scientific academy, 28–29; and mathematics, 25–27; measuring, 28, 129, 156n34; microscope, 14, 27; mirrors as, 45; musical, 113; and observation, viii, 17, 29, 86, 140n86; optical, 24; in the Orphanage, 1, 6–7, 57, 126, 142n25; and philanthropy, 111, 113; of reason, 22–23; in Ritter academies, 16; scientific, 111, 156n19, 162n79; Semler's construction and use of, 63, 83, 106; Solomon's Temple models as, 62, 66–67; and vocational training in Zittau, 115–16
irenicism, 37–44, 50, 54, 96, 128
isoclinic maps. *See under* map(s)

Jablonski, Daniel Ernst, 38, 40–43, 55, 147n22
Jesuits: colleges and schools, 15, 32, 91, 121; Kircher, 93, 98–99, 156n34; and Noël's magnetic observations, 93, 96, 98–99; and scientific missions in China, 31, 33, 96; and spiritual exercise, 40, 44

Kircher, Athanasius. *See under* Jesuits

Langendorf Orphanage, 19, 108–9, 111–16
LaVopa, Anthony, 9, 137n31
Leibniz, Gottfried Wilhelm: and affect, 11–13; as friend of Francke's, 10, 12; and mechanistic worldview, 11; on observation, viii
longitude problem, 7, 19, 87, 92–99, 103–7, 154n3
love: as awakening, 74, 77; and emotional intelligence, 8, 11–13, 127; of mathematics, 27, 63; and mechanical philosophy, 48–49, 53; through observation, 45–47, 61, 129; as philanthropy, 11–12, 74, 130, 137n42; and Pietism, 8, 39, 45, 136n22; potion, 88; seraphic, 48, 113; of truth, 128–29
ludus, learning as, 25–26, 37, 63, 129, 142n35